# IMechE Engineers
# Careers Guide

# IMechE Engineers' Careers Guide

## Abby Evans

A John Wiley and Sons, Ltd, Publication

This edition first published 2010

© 2010 John Wiley & Sons Ltd

*Registered office*
John Wiley & Sons Ltd, The Atrium, Southern Gate, Chichester, West Sussex, PO19 8SQ, United Kingdom

For details of our global editorial offices, for customer services and for information about how to apply for permission to reuse the copyright material in this book please see our website at www.wiley.com.

Reprinted April 2011

**Some sections of this book draw on the careers resources developed by the Association of Graduate Careers Advisory Services (AGCAS) that are published on the Prospects graduate careers website. Thanks to AGCAS who gave permission for the material to be used in this way.**

*Library of Congress Cataloging-in-Publication Data*
Evans, Abby.
  IMechE engineers' careers guide / Abby Evans.
     p. cm.
  ISBN 978-0-470-97419-3 (pbk.)
  1. Mechanical engineering--Great Britain--Vocational guidance. I. Title.
  TJ157.E93 2010
  621.023'41--dc22

                                                                                          2010024887

ISBN 9780470974193

A catalogue record for this book is available from the British Library.

Set in 10.5/14.5 pt Palatino by Sparks – www.sparkspublishing.com
Printed in the UK by TJ International Ltd, Padstow, Cornwall

# Contents

# Foreword

This book is a planning tool to help you as you embark on your mechanical engineering career, providing orientation on the industry and acting as a starting point in the job hunting process. With insights into a range of sectors and information on the industry as a whole, this book will serve as a guide to professional success. With crucial advice on the key stages of getting a job, and guidance on professional qualifications and training once in employment, this book will answer questions about the engineering industry and enable you to plan for your career.

In a competitive, fast paced and demanding industry, newly graduating engineers need to understand the opportunities available to prepare for employment and to ensure that they are constantly learning and advancing their knowledge and skills. The Institution of Mechanical Engineers (IMechE) is here to help you achieve this and we are committed to providing the best possible service for our members. Home to 94,000 engineering professionals working in the heart of the country's most important and dynamic industries, we will ensure that you have the skills, knowledge, support and development advice you need at every stage of your career.

Our members enjoy industry recognition for their professional competence and enhanced career opportunities as a result, because we set the standard for professional excellence in mechanical engineering. To achieve this recognition of your skills and to manage your career development, it's important that you maintain your membership of IMechE and take advantage of the opportunities we have to offer to help you fulfil your potential.

Upon graduation from your degree you can apply to become an Associate of IMechE and begin the journey towards professional registration. With appropriate professional work experience and support from IMechE to develop your skills and knowledge, you can apply for registration as an Incorporated or Chartered Engineer, to mark you out as a member of the engineering elite and take your career from strength to strength.

Your membership of IMechE will bring ongoing support for your continued professional development, through a range of member resources, events and benefits. Engineers need to continue their professional development to keep their skills fresh and progressive, so we'll help you stay up to date, broaden your knowledge and deepen your understanding of your chosen industry.

We hope that your relationship with IMechE will be a lifelong one that supports you throughout your career. As you join this exciting and essential profession, we wish you luck and look forward to helping you stay ahead in an increasingly varied, dynamic and rewarding industry.

# About the Author

Abby Evans is a Careers Adviser at Oxford University where she specializes in supporting students interested in engineering and science careers. On completing her degree in Earth Sciences at Cambridge University she continued her studies there to complete a PhD in Marine Geophysics in 1995. After a PGCE at Leicester University and four years as a science teacher in a Leicester comprehensive school, Abby began work as a careers adviser at Oxford in 2001. She has written and contributed to many national publications on careers in engineering, science and education, as well as developing a series of careers publications for Oxford University students.

# Acknowledgements

IMechE would like to thank Abby Evans for compiling this book. Some sections of this book draw on the careers resources developed by the Association of Graduate Careers Advisory Services (AGCAS) that are published on the Prospects graduate careers website. Thanks also to AGCAS who gave permission for the material to be used in this way.

IMechE would like to thank the following members for their contributions to editing the industry chapters of this book:

Group Captain Timothy Brandt, CEng FIMechE
James R. W. Bridge, AMIMechE
Prof. Antony M. J. Bull, CEng FIMechE
Ian P. Burdon, CEng FIET
Dr Patrick A. Finlay, CEng FIMechE
Squadron Leader Paul Lloyd, CEng MIMechE
Stephen Phillips, AMIMechE
Mike K. Rolls, CEng FIMechE
Andy M. Teague, CEng FIMechE
Oliver Tomlin, CEng MIMechE

# 1
# Engineering Industry Overview

## 1.1 Engineering in the UK

The engineering sector is huge. In the UK alone it employs over 4.5 million people and contributes almost a fifth of UK GDP. From 2008 to 2009 the number of UK engineering enterprises increased by over 12,000 to 482,880, with the largest increases in London and the South East of England (source: Engineering UK 2011 report). Today, many engineering sectors are thriving. The UK manufacturing sector is the sixth largest in the world and accounts for three-quarters of all industrial research and development in the UK.

The UK is host to some of the world's top engineering enterprises. All sectors of industry are represented, from aerospace and automotive through chemicals and energy to manufacturing, transport and utilities.

The **energy sector** in the UK has always had its share of big players. It includes the oil and gas companies plus the electricity power utilities that are developing new low-carbon technologies such as biogas, landfill gas, hydrogen, solar, wind, wave, etc. Higher prices have enabled the UK off-shore oil and gas industry to exploit previously uneconomic deposits off the UK coast. Power generators are planning massive campaigns to build renewable and nuclear power generation plants. These developments have resulted in more jobs and opportunities in the industry, especially for graduates in engineering.

The UK's **aerospace** industry is the largest in the world outside the USA. It is a significant driver of regional and national economic growth and productivity. In the UK **defence** equipment market, products include military aircraft, satellites, rockets and missiles, navigation and electronic guidance systems. Hundreds of small firms act as suppliers to the industry and there is work for a broad range of engineers from disciplines including aeronautical, structural, electronic and mechanical engineering.

The UK has a strong **manufacturing** base, employing 2.8 million people, and with a turnover of £502 billion in 2009. However the trend from high volume, low value manufacturing to more specialist products has meant that, although turnover has increased in recent years, the number of employees in manufacturing has declined.

The UK has an extensive **transport** network of around 400,000 km of roads carrying an estimated 34 million vehicles, a rail network that carries 1.2 billion passenger journeys per year and a fleet of naval, merchant and passenger ships. Add into this the light rail, tram and underground networks, aviation and freight operators, and the scale of the transport sector soon becomes clear.

Opportunities in the **automotive industry** are more prone to peaks and troughs than many other industries and positions for engineering graduates can be hard to find. However, the UK boasts several major car manufacturers as well as a number of smaller producers serving specialist markets, such as sports and luxury cars and London taxis. There are, in addition, over 1,000 automotive component suppliers manufacturing in the UK, 90% of which are SMEs (small to medium sized enterprises), with many offering high levels of expertise in specific technical areas. The UK is recognized as a world leader in innovation in component manufacture and attracts considerable investment from international manufacturers. It is also the centre of the motorsport world. Motorsport is one of the UK's major export earners with a total turnover of £1.3 billion.

## 1.2 Is There a Shortage of Engineers?

Educators and industrialists agree that schools in the UK are not producing enough pupils who have studied physics and maths at A-level or Scottish Higher level. The study of these subjects at advanced level is crucial for entry into engineering degrees.

The Confederation of British Industry (CBI) states that the UK will need to double the proportion of science and engineering graduates leaving university by 2014 or see skilled jobs go overseas (CBI, 2007).

In some areas of engineering, the demand for well-qualified, skilled graduates outstrips supply, and some recruiters find that they have hard-to-fill vacancies. However, the Association of Graduate Recruiters (AGR) reported a fall of 8.9% in the overall number of graduate vacancies in the UK in 2009, and graduate unemployment at a 17-year high in 2010. More recently, increases in the number of graduate opportunities have been reported by the Chartered Institute of Professional Development and by the AGR in the 2010–2011 recruitment season, with an overall predicted increase in the number of graduate vacancies of 3.8% compared to 2009–10.

## 1.3 What Do Engineering Graduates Do?

The engineering sector offers you the opportunity to travel, work on exciting and innovative projects, and get your hands dirty with practical work and responsibility early on in your career. It encompasses a huge range of occupations, including:

- product and process development;
- manufacturing;

- consultancy;
- research and development;
- design, construction, commissioning and operation;
- data management;
- IT support;
- logistics;
- management and administration;
- sales.

Opportunities exist in major towns and cities throughout the UK. Some regions have traditionally hosted particular industries. For example, companies working in high-tech areas of engineering are most prevalent in the 'golden triangle' of Oxford, Cambridge and London, and there is a concentration of automotive companies in the West Midlands. Many companies are multinationals offering opportunities to spend periods of time working overseas on short-term projects, or for extended periods of time. For example, although the UK oil industry extends throughout the UK (albeit with a particular focus on Aberdeen), oil exploration has always been international in nature, with seismic investigations and the drilling of wells often taking place in the most remote regions. The industry is used to working in multinational teams, and English is the business language.

The international nature of many areas of engineering is further highlighted by the transferability of professional engineering status across the European Union (EU). All EU countries have an agreement on what they call the 'formation' of a professional engineer, which takes around seven years and includes education, initial training and experience. The order in which these occur differs around Europe. The French and Germans prefer their students to gain industrial experience at home and abroad during the course of their studies. In contrast, in the UK, although we have sandwich courses, work experience before graduation is more restricted and it is quite possible to graduate without any. This inevitably places more emphasis on industrial training after graduation.

It is relatively difficult for a new graduate to secure a permanent job overseas (EU nationals wishing to work in EU countries are an exception). If you are keen to work abroad it may be worth looking for employment in a multinational company where there could be opportunities for overseas postings. Once you have gained a few years of experience then the international job market is likely to open up considerably.

According to the analysis of graduate destinations statistics by EngineeringUK there is a positive employment picture for engineering and technology graduates (Figure 1.1). Fifty-nine per cent of UK domiciled engineering and technology graduates are in full-time paid work six months after graduating. Almost three-quarters of engineering and technology graduates who enter employment work for employers in sectors in or associated with engineering and technology (Figure 1.2).

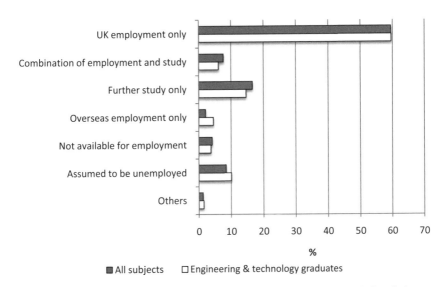

**Figure 1.1** Destinations six months after graduation of those completing first degrees at UK universities in 2008 (data source: HESA)

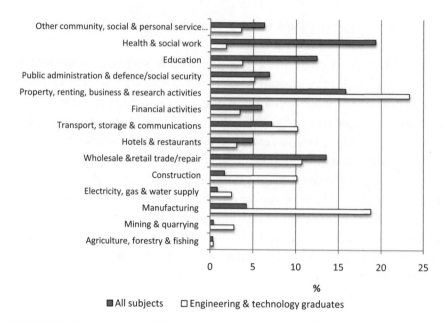

**Figure 1.2** Industry sector of those in employment six months after graduation with first degrees from UK universities in 2008 (data source: HESA)

## 1.4 What Is It Like Working in the Engineering Sector?

**Working conditions** vary according to your role. You could work in an office, in a laboratory, on the factory floor, on site, outdoors – in any number of combinations. Some roles – for example, site engineer or field engineer – may require extended periods of time working away from home. Others require regular office hours, or perhaps shift work.

**Salaries** for newly graduated trainee engineers compare favourably with pay for other graduate jobs according to the CBI's Education and Skills Survey 2009. The median salary of a graduate engineer is £22,500. A recent longitudinal study of graduate engineers shows that their salaries 3.5 years after graduating were third highest (at £28k, after medicine and dentistry, and veterinary science) of all subject areas. The average for all subjects at this point was £24k.

**Women** are under-represented in the engineering industry. Analysis of new registered engineers and technicians, undertaken by EngineeringUK, shows that in 2009 just over 12% of registrants were female. This gender imbalance does not begin in the workplace however. The proportion of women studying engineering and technology in higher education institutions is just above 15%. This compares poorly even with other technical subjects – in biological sciences 64% of students are female and in physical and mathematical sciences it is 41% and 38% respectively. The careers followed by graduates on leaving education are also highly gender-orientated. Issues arising from this under-representation are widely acknowledged. Engineering companies and professional institutions are taking steps to understand the underlying causes and try to shift the balance. The Institution of Mechanical Engineers, for example, issued a policy statement *Gender in Engineering* (IMechE, Education Policy Statement 04/09) in which recommendations were made not only to government but also to schools and to engineering employers. Several large companies and professional organizations have mentoring programmes for women, and sponsorship schemes. There are also a number of projects and campaigns set up to encourage women into engineering, for example, Women Into Science, Engineering and Construction (WISE). There are many inspiring and reassuring case studies of women who have pursued highly successful careers in engineering. For examples consult the IMechE Engeneration Network, the WISE website and NOISE blogs.

**Consultancy versus in-house:** One feature of the engineering industry is that many companies use external consultants to provide knowledge or skills rather than developing that expertise themselves in-house. This includes consultants who are employed by firms which specialize in consultancy services, as well as individuals who run their own small businesses capitalizing on their own expertise. Recent graduates will find it difficult to work as an individual, or freelance, consultant before gaining significant experience in their chosen industry. There are, however, many engineering consultancy firms who take on graduate engineers each year. These may be global multidisciplinary companies who provide expert services and technology to a range of engineering industries on a contractual basis. Examples of these types of organizations include Atkins, Scott Wilson and AMEC. Other consultancies such as Newton Consulting work on short-term projects to troubleshoot manufacturing processes or improve industrial productivity. Advantages of consultancy work include the variety of projects and exposure to a range of businesses and industries, often involving travel, and providing valuable experience for future employment. However, for some the need to market the services of your firm, and liaise extensively with clients, whilst useful business experience, can feel like a distraction from their vocation.

## 1.5 How Can my Career Develop?

Your career as an engineer could develop in many different ways. You might start in a technical function and develop as a professional, before moving into a managerial position. This may involve managing a facility such as manufacturing or a project for a client. In addition to your technical abilities, you will also develop the skills necessary to manage people, budgets and clients' expectations.

Gaining the professional qualification of Chartered Engineer (CEng) or Incorporated Engineer (IEng) is the next step for many graduate engineers. The Engineering Council, the UK regulatory body for the engineering profession, holds the national register of 235,000 Chartered Engineers, Incorporated Engineers and Engineering Technicians in the UK. Such professional titles are gained by demonstrating that you have built up the necessary competencies and experience in your job as an engineer. IMechE offers an Initial Professional Development (IPD) programme that will build on your academic skills and lead you to registration as a Chartered Engineer (CEng) or Incorporated Engineer (IEng). You are likely to be supported in this process by your employer. These qualifications show that engineers have achieved a benchmark level of competence and a commitment to continuing professional development in engineering. Many senior-level engineering posts require

Chartered status. You can read more about professional qualifications in engineering in Chapter 8.

Alternatively, your career could progress in a different direction, perhaps marketing the products you know so well technically, negotiating with suppliers in a procurement capacity, optimizing supply chains or hiring new technical staff in human resources.

## 1.6 Future Trends for the Engineering Sector

Engineering is an ever-changing industry. We may not know the future with certainty but there are clear signs of developments to come. Engineering has featured in a number of government policy initiatives in acknowledgement of the importance of these sectors to the country's economic future – especially during a recession. In particular, the sector looks set to benefit from government support for power generation, low carbon technologies and other advanced engineering projects. In April 2009, the Department for Business, Enterprise and Regulatory Reform (BERR) published *Building Britain's Future: New Industry, New Jobs*, in which it identified several key technologies that should play a larger role in the economy in the future, namely advanced engineering, electronics, biosciences and low-carbon technologies.

### 1.6.1 Sustainable Energy and Low Carbon Industries

The most important issue facing the engineering sector is thought by many to be the need to safeguard the environment. Engineers are at the forefront of developing innovative technology to allow for sustainable development, for example through the reduction in carbon emissions by industry and the development of alternative sources of energy to reduce our reliance on fossil fuels. Environmental engineering is a relatively new profession but is already increasing in importance. There are opportunities in this area for mechanical engineers in contributing to the next generations of renewable energy sources, improving methods and efficiencies of existing industrial processes, involvement in the next phase of the nuclear power programme, developing more efficient and environmentally friendly transport systems, and so on.

More and more engineers, regardless of their discipline, are involved in projects concerned in some way with the problem of energy. A combination of rising energy prices, fears over potential disruption to power supplies in the coming years and mounting evidence that the UK is not on course to meet the government's carbon emissions targets all mean that the subject of energy is currently never far from the headlines.

The immediate challenges facing engineers are to reduce emissions from carbon-based fossil fuels, to identify alternative sources of energy and feedstock, and to help manage the economic transfer of dependency from one source to another.

Replacing traditional nuclear energy with nuclear fusion offers the potential for limitless, environmentally friendly energy. However, this potential is unlikely to be realized for some time and until then engineers will have to come up with ways of extending the lives of the nuclear power stations and developing alternative sources of energy.

The development of a low carbon industrial strategy was identified as a key step for the future in the government paper *Building Britain's Future: New Industry, New Jobs* (HM Government, April 2009).

### 1.6.2 Nanotechnology

Nanotechnology promises to have a huge impact on engineering with applications in many industries, but especially in medicine. It will provide earlier and better diagnostics, and more precisely targeted drug delivery for treatments.

For example, nanotechnology, in the form of flexible films containing miniaturized electrodes, is expected to improve the performance of retinal, cochlear and neural implants. It could also lead to the miniaturization of medical diagnostic and sensing tools.

In could be that nanotechnology could enable developing nations to leapfrog older technologies, in the way that copper wire and optical fibre telephony were superseded by mobile phones.

Another feature of nanotechnology is that it is an area of research and development that is truly multidisciplinary. Research at the nanoscale is unified by the need to share knowledge on tools and techniques, as well as information on the physics affecting atomic and molecular interactions. Materials scientists, mechanical and electronic engineers, and medical researchers are now forming teams with biologists, physicists and chemists.

## 1.7 Further Resources

- Women into Science Engineering and construction (WISE)
      http://www.wisecampaign.org.uk/
- Institution of Mechanical Engineers Engeneration
      http://www.engeneration.imeche.org/
- New Outlooks In Science and Engineering (NOISE)
      http://www.epsrc.ac.uk/noise
- Engineering Council
      http://www.engc.org.uk/
- Higher Education Statistics Agency
      http://www.hesa.ac.uk/
- What do graduates do?
      http://www.prospects.ac.uk/links/wdgd
- EngineeringUK
      http://www.engineeringuk.com/
- Confederation of British Industry (CBI)
      http://www.cbi.org.uk
- Association of Graduate Recruiters (AGR)
      http://www.agr.org.uk/

# 2
# Energy and Utilities

**Figure 2.1**  Electricity transmission tower, near Aust, UK. Taken from: http://
en.wikipedia.org/wiki/File:Pylon_ds.jpg. Source: Yummifruitbat

---

> **At a Glance**
>
> - A thriving industry with engineering at its heart, with the challenge of rapid technological advancement to develop secure and sustainable energy resources. Engineers working on offshore rigs, developing innovative technologies, operating and maintaining power stations, supporting and developing a safe nuclear industry.
> - Employers: global oil companies, national and international energy companies, companies providing specialist services, expertise and technologies.
> - Starting salaries: £25k–£30k in major oil companies, £22k–£28k typical elsewhere.

## 2.1 Overview

The energy sector in the UK directly employs around 135,000 people and many more jobs support the industry. It is an area that continues to make rapid technological advances, and is one of the few sectors that continued to flourish during the recent times of economic downturn. A great deal of emphasis is placed on the need to develop secure and sustainable energy supplies. At the heart of many areas of the industry is the drive to make existing supplies last longer as well as developing new ones.

The extraction of oil and gas at its current rate is not sustainable in the long term. Although innovations in extraction methods, and rising prices, combine to make less accessible reservoirs commercially viable, there will come a time when our dependence on fossil fuels will have to end. New technologies, particularly in the field of renewable energy, are already making an impact on the energy sector. The UK Government, for example, has recently issued licences for the construction of nine new major wind farms off the coast of the UK that could generate a quarter of the UK's electricity needs.

Such developments and the challenges associated with reducing our dependence on fossil fuels make this an exciting time to join the energy industry. Bear in mind though, that in the short term at least, the most plentiful career opportunities remain in the oil and gas industry, and that the majority of our electricity still comes from burning fossil fuels.

According to the Organization of Petroleum Exporting Countries (OPEC) more than 80% of the world's energy requirements still come from oil at the present time.

The future for the energy sector looks buoyant. The International Energy Agency (IEA) has forecast that world energy demand is likely to rise by over 50% between now and 2030. Energy prices are expected to rise in the next few years in line with increasing demand.

---

Roles for mechanical engineers in the energy and utilities sector:

- engineer in the oil and gas industry (this chapter);
- engineer in power generation and distribution (this chapter);
- engineer in the renewables sector (this chapter);
- water engineer (this chapter);
- process engineer (Chapter 6).

---

## 2.2 Typical Employers

Many of the big players in the energy sector describe themselves as **integrated energy companies**. By this they mean that they use a range of fuel sources (oil, gas, coal, nuclear and renewables) to generate the electricity that they then distribute to homes and industry. These companies have evolved from the energy utilities companies of the past. Such organizations were typically regional operations and this heritage is sometimes still reflected in the name of the company – Scottish and Southern Energy, for example. However, most now operate nationally and even internationally. Examples include Centrica, E.ON, EDF Energy and npower.

**Fossils fuels** account for around ten times the energy produced from all other sources (MacKay, D. 2008; *Sustainable energy: Without the hot air*, UIT, Cambridge). Oil is the single most-used source of energy, and there is a huge global industry involved in oil and gas exploration, extraction and production. Global players with opportunities for mechanical engineering graduates include oil companies such as BP, Shell, ExxonMobil, etc.

Oilfield service companies such as Schlumberger and Halliburton provide specialist services to oil companies to aid the exploration and extraction of oil but do not directly extract oil and gas themselves. The UK's oil and gas industry has been self-sufficient since 1980, but is not able to retain self-sufficiency for a great deal longer as the UK's oil and gas reserves decline (Cogent: The Sector Skills Council for Chemicals, Pharmaceuticals, Nuclear, Oil and Gas, Petroleum and Polymers). In the UK much of the oil industry is based around Aberdeen and Great Yarmouth.

The **nuclear industry** in the UK provides about 18% of the UK's electricity. The industry consists of five main sectors: (1) power generation; (2) decommissioning; (3) processing and reprocessing nuclear fuel; (4) new build; and (5) defence. The roles of most relevance to mechanical engineering graduates are those involved with power generation. The nuclear industry looks set to play an increasingly important role in the UK energy mix following recent government backing for a new generation of nuclear power stations.

The ten nuclear power stations currently generating electricity in the UK are run by companies such as British Energy Generation (part of the EDF Group) and Magnox. Other companies, such as Sellafield Ltd, are involved in decommissioning plants and waste management. The Nuclear Decommissioning Authority (NDA) has overall responsibility for the UK's civil public sector nuclear assets and for contracting out the decommissioning and clean-up of the UK's civil public sector nuclear sites. The industry is supported by a number of nuclear technology services providers such as the National Nuclear Laboratory and Westinghouse. Such companies provide specialist expertise to the nuclear industry across a huge range of services from the development of technology to the design and manufacture of reactor systems to advice on waste management and decommissioning.

The **renewable energy** sector covers a wide range of energy sources and technologies including solar, wind, wave, tidal, biofuel and geothermal. In the UK the renewables market is dominated by wind energy, with hydro-electricity and biofuels also becoming established. Tidal power is less well-developed in the UK, although the Bristol Channel presents a site with great potential. Solar power has been relatively slow to develop in the UK. The type of company, and opportunities for graduate mechanical engineers, varies greatly according to the type of renewable energy they work with. Typical employers in the renewables sector include companies that develop technology and manufacture equipment, energy companies and utilities companies.

Organizations vary enormously both in size and degree of specialism. International energy companies such as BP and Shell carry out research and development into a wide range of renewable technologies. Other international companies specialize in a single energy source – for example Vestas, which works with wind turbine technology. Utilities companies, which may have a national or international presence, are increasingly adding renewable sources of energy to their mix of electricity generation methods. Examples in the UK include npower, Centrica and Scottish Power. At the opposite end of the scale are a myriad of small specialist companies that focus on developing and marketing a specific renewable technology. Other employers of engineers in the renewables sector include research departments in universities.

Although most engineering opportunities within the **water industry** are for civil and environmental engineers, some roles are suitable for graduates with a background in mechanical engineering. In the UK ten regional water companies handle both the supply of clean water and the disposal of waste water, supplemented by 15 smaller companies dealing in clean water only. Larger companies have diversified into overseas work in addition to their statutory duties on the domestic front. Some of the largest water companies in England and Wales, particularly those that supply an entire region, have become part of larger multi-utility groups that also include gas and electricity companies. These groups may be international providers of utilities to many countries.

The position in Scotland and Northern Ireland is different, with one public sector body responsible for water matters in each country. A complete list of UK water companies can be found on the website of Water

UK. Large construction firms may have water engineering departments (sometimes known as environment) which work with other departments, such as highways or construction, and give the opportunity to develop skills and knowledge in different engineering disciplines. The main government agencies that employ water engineers are the Environment Agency (EA) in England and Wales, and the Scottish Environment Protection Agency (SEPA). British Waterways employs water engineers to maintain its extensive network of canals and associated storage dams. Local authorities also have some opportunities for water engineers, although generally on smaller local watercourses and coastal and river defences.

## 2.3 Engineering Roles Specific to the Energy and Utilities Sector

As an energy engineer you could be involved in the production of energy from fossils fuels and nuclear sources as well as from renewable or sustainable sources, such as wind, wave, tidal, biofuels and solar power. You might work in a variety of roles including designing and testing machinery as well as developing ways of improving existing processes to reduce cost and minimize environmental impact.

In the oil and gas industry you might be involved in designing, developing or operating machinery used on offshore installations to extract natural oil and gas from many metres beneath the seabed. You might be involved in the conversion of crude oil in its raw state to the many types of hydrocarbons that the world is so dependent on by working in a refinery ensuring the safety of these volatile products. Alternatively the future of natural gas lies in liquified natural gas, or LNG. This converts gas into liquid form for ease of transportation. The use of cryogenic procedures reduces the volume of the gas by 600 times.

In the renewables sector you could be involved in the design and development of wind turbines. Alternatively you may be a nuclear engineer, ensuring that plants run productively and safely, or even supervising the construction and maintenance of Royal Navy submarines (Chapter 3).

## 2.3.1 Energy Engineer – Roles in the Oil and Gas Industry

The oil and gas industry is commonly split into two main components. Upstream refers to the searching for and recovery of oil and gas – often termed exploration and production. Downstream refers to the refining of crude oil, and distribution of its products and of natural gas. The downstream sector includes oil refineries, petrochemical plants, distribution networks and retail outlets.

Downstream roles with a mechanical engineering focus are most commonly found in refineries where engineers are responsible for developing, operating and maintaining the complex equipment required to refine crude oil into more useful and commercial components. The role of process engineer is covered in Chapter 6.

In the upstream sector mechanical engineers play a vital role in developing, operating and maintaining the complex equipment required to find and produce oil. You may be involved in projects at the development phase of a project dealing with rotating machinery, static equipment or mechanical systems, ensuring that the correct equipment is specified and is fit for purpose; or involved in maintaining existing facilities, developing and implementing maintenance programmes, and troubleshooting operating problems. There are many terms for the different upstream roles for mechanical engineers. The main ones you are likely to come across are production engineer, well engineer and drilling engineer. Here we will take a closer look at the role of **drilling engineer** to get a flavour of reality of an engineering role in the upstream oil and gas industry.

A drilling engineer develops, plans, costs, schedules and supervises the operations necessary in the process of drilling oil and gas wells. They are involved from initial well design to testing, completion and abandonment. Engineers are employed on land, on offshore platforms or on mobile drilling units either by the operating oil company, a specialist drilling contractor or a service company. The role can involve administering drilling and service contracts, engineering design, the planning of wells and supervising the drilling crew on site. Drilling engineers work with other professionals, such as geologists and geoscientists, to monitor drilling progress, oversee safety management and ensure the protection of the environment.

Typical work activities for a drilling engineer include designing and selecting well-head equipment and drawing up drilling programmes, taking account of desired production flow rates. In the operation of the well, a drilling engineer monitors the daily progress of well operations and current daily costs, comparing actual costs with cost expenditure proposals and recommending changes or improvements to rig work techniques, which could lead to optimization of expenditure. They also carry out analysis on site and recommending immediate actions as necessary.

Often you will be required to work offshore or in remote areas, but office-based roles are available too. In the North Sea, offshore working hours are normally 12 hours on and 12 hours off continuously for two weeks, followed by a break ashore of two to three weeks. You may be summoned to work on a rig at short notice.

Oil exploration is an international activity and the work of a drilling engineer can take you all over the world, from Africa to Eastern Europe to the Middle East. If you work overseas you might work 'on rotation', spending up to two months onsite followed by a break of one month at home. The work on offshore rigs is demanding and physically hard, often undertaken in dirty, wet and noisy conditions. The weather may also be unpleasant. Living conditions on most rigs are very comfortable – sometimes described as five-star hotel standard quality. Rigs usually accommodate 50–100 people. Rooms are compact but comfortable. They are frequently shared with a colleague working the opposite 12-hour shift, so you will rarely see your room-mate. All meals and laundry services are provided. Regular facilities include a gym and snooker room with access to computers and DVDs. Alcohol is prohibited on the rig.

Field and offshore operations take place in many parts of the world (high seas, remote jungles, vast deserts and mountain ranges). In some settings, you may be working for several months at a time without a break. You will need to be prepared to accept considerable disruption to your personal and home life. Communal living with the same set of people means you need to get on well with others, both in and outside the work situation.

The range of typical starting salaries for drilling engineers is £29k–£33k. These numbers relate to international oil company graduate training programmes, with the highest figure available to holders of a relevant PhD. Salaries in smaller companies are likely to be lower. Salary is usually performance-related. Location and assignments influence salary. Additional fringe benefits and overseas allowances may be available. Oil drilling takes place in some dangerous areas of the world. In these situations, extra payments may be made.

## Getting In: Entry

A relevant pre-entry postgraduate qualification can be useful but is not a guarantee of a job due to the high level of competition. Any student considering postgraduate study should investigate carefully the need for and relevance of the proposed course to their intended career. An MSc in petroleum or offshore engineering, for example, may improve your chances.

Pre-entry experience is not essential, but any experience working on rigs or in an onshore yard is likely to be useful and demonstrate your interest and motivation. Some major oil companies and contractors offer placements during the summer vacation of your penultimate year where you can work on a project of operational significance. Successful completion of the project may give you an advantage in the recruitment process for a permanent post.

Major oil companies advertise vacancies almost a year in advance, but it is also worth making speculative applications to specialist companies directly. Competition is keen, and recruitment is affected by oil price fluctuations.

## Getting On: Career Prospects

If you begin your drilling engineer career with one of the large oil companies, you may initially manage a single well under supervision. However, fairly quickly, you could become responsible for wells involving budgets of £5–£10 million. As you gain in experience and seniority, you could take on the overall supervision for the drilling and production operations on several wells – initially offshore and then moving onshore.

Formal training in areas specific to the job, as well as more general management skills, complements practical, hands-on, rig-site experience. Major oil companies have well-established graduate training programmes. These programmes help drilling engineers meet the requirements of the professional engineering institutions in order to gain Chartered Engineer (CEng) status through bodies such as the Institution of Mechanical Engineers (IMechE). A mentoring system is often available to allow new graduates to access the advice and professional support of more senior engineers. Continuing professional development (CPD) is supported at a variety of locations – allowing graduates to build networks with international colleagues. Study facilities are also available offshore on oil platforms. If you are working for a smaller company, you may find that you need to take responsibility for arranging and funding your own development and training, particularly if you are employed on a temporary contract basis.

A typical career path could involve working for two to four years offshore or on an onshore well site and then moving into an office-based design role. This path might eventually lead to working in an overseas office following a one month on, one month off pattern.

Career progression in oil companies is usually into management. However, with drilling contractor work, engineers tend to remain in a technical role, using their expertise to access and develop the most appropriate technology for drilling in the future. Independent consultancy is another option, although you will need to work hard to balance the good times with the less prosperous periods.

Drilling is the aspect of oil exploration most affected by the economic climate. Changes in the economic climate may affect opportunities for career development. Activity is in response to the decisions of the oil companies as to where drilling will take place. North Sea sites are currently being transferred to smaller developers as the major fields become less economic. Large companies are increasingly looking to new prospects overseas.

The range of typical salaries after 15–20 years in the role is £60k–£100k+.

## 2.3.2 Energy Engineer – Power Generation and Distribution

The power generation sector uses an incredibly wide range of technologies to provide essential electricity to homes, businesses and industry. Technologies range from well-established steam turbine systems driven by the burning of fossil fuels or heat generated in a nuclear reactor, to pioneering technologies that harness sustainable sources of energy such as wind, wave and solar. Roles for engineers in the renewables sector are covered in the next section.

There are many different engineering roles available in the power generation and distribution sector. You could be involved in improving turbine design for a longer working life and lower fuel consumption or developing distribution networks.

Operational roles based in power stations include control room operations, plant safety systems, routine testing and environmental impact compliance. You may be involved in planning and carrying out overhauls of power stations, or managing projects to install new power plants. You might be using advanced inspection and modelling techniques to investigate reasons for failure of steam turbines and to engineer speedy and innovative solutions.

Working hours vary according to the area of the industry. Design, research and development roles generally work from nine to five, Monday to Friday. However, in power plant operations, hours may be based around a seven-day rolling shift system to provide cover 24 hours a day, seven days a week. Some staff may be on standby for call-out in case of emergencies.

Power station environments can vary from clean area work spaces such as control rooms, to workshop-type environments. Power plants can be hazardous; health and safety procedures are strictly followed and given high priority.

Starting salaries for new engineering graduate are typically in the range £24–£26k.

## Getting In: Entry

The demand for high-quality graduate applicants across the energy industries looks set to continue for the foreseeable future. Most employers have comprehensive graduate recruitment web pages containing a wealth of information about graduate opportunities. Employers visit universities to give presentations and attend careers fairs, usually in the autumn term. Most graduate training schemes are advertised during the autumn term for the following September start.

Although work experience in the sector is not essential, a number of organizations offer summer courses, one-year industry schemes and summer internships, enabling undergraduates to obtain a real insight into the industry. Apart from advertised work experience programmes and placements, work experience may be found by making direct speculative approaches and using contacts, alumni associations and past students – see Chapter 7 for advice on finding work experience.

## Getting On: Career Prospects

Most large firms offer structured training and encourage professional development. Usually, firms offer in-service training and short courses for specific needs. Your in-house training may include placements in different departments to broaden your experience.

To progress in the profession, it is becoming increasingly important to achieve professional status as a Chartered Engineer (CEng). The Engineering Council UK (ECUK), in collaboration with the main engineering institutions, has introduced regulations known as UK Standard for Professional Engineering Competence (UK-SPEC). These affect all graduates aspiring to achieve CEng status. As companies operate in an increasingly international market, the European engineer (Eur Ing) status and additional language skills will become a distinct advantage if you wish to progress further in the profession. All Chartered Engineers are eligible to apply for Eur Ing status.

Some mechanical engineers move into business functions, such as procurement, sales and marketing, or human resources. Others will develop their technical skills and look to move into senior engineering posts, such as engineering director. Another career path could involve moving into engineering consultancy.

### 2.3.3 Mechanical Engineer in the Renewable Energy Sector

Energy has now become a topic of huge importance and interest to individuals, industry and government. Climate change and the increased price of oil and gas as an energy source have prompted the government to make sustainable and renewable energy generation a priority. The 2007 Energy White Paper, Climate Change Act 2008 and the UK Low Carbon Transition Plan 2009 are all focused on reducing carbon emissions. Consequently, the renewable energy industry is expanding rapidly. The demand for oil and gas is increasing, and pressures for businesses to reduce carbon emissions and be more energy efficient has led to a growth in renewable or sustainable sources of energy such as solar, wind and hydropower.

About a fifth of those employed in the renewables sector are in technical roles. Mechanical engineers working in the renewable energy sector are focused on finding efficient, clean and innovative ways to supply energy. They work in a variety of roles including designing and testing machinery, developing ways of improving existing processes and converting, transmitting and supplying useful energy to meet our needs for electricity. The nature of the role obviously depends on the type of energy – for example, the majority of wind projects are in construction.

Starting salaries vary enormously. A new graduate working for a large energy or utility company typically receives £22k–£30k. First posts in smaller, specialist companies are likely to command slightly lower salaries than this.

Working hours vary according to the area of the industry. Design, research and development roles generally work from nine to five, Monday to Friday. In power plant operations, hours may be based around a seven-day shift system. The work can be office, laboratory or site based, with site visits and fieldwork being conducted outside in all weathers. Some jobs may be offshore or in remote and isolated locations.

There are opportunities across the UK. Many companies are based in London and the South East, but it is common to find companies based in the regions where the energy source they exploit is most plentiful. Scotland and the South West of England, therefore, are home to large numbers of wind energy companies.

## Getting In: Entry

Many graduates are interested in a career in the renewable energy sector, but opportunities for new graduates can be difficult to find. There are many large companies that offer graduate training schemes in the energy sector – but virtually none will guarantee that a new graduate will spend their time working exclusively in renewables. It is more likely that you will spend time working on a range of projects in more mainstream energy technologies involving fossil fuels before opportunities to gain experience and specialize in the comparatively small area of renewables are possible. One of the few graduate schemes that is purely focused on renewables is offered by Vestas, an international wind power company that recruits 60–80 graduates onto its scheme each year.

Most companies that operate solely within the renewables arena have relatively few opportunities for new graduates with little or no experience of the sector. Building up your experience of the sector is therefore vital, and a great deal of research into potential employers, speculative approaches to companies and persistence may be necessary.

One way to develop expertise and gain some experience of the sector is by taking a relevant postgraduate course. As renewable and sustainable energy solutions are increasing in both their profile and application, so too is the number of specialist degrees that focus on energy and the environment, such as energy engineering, sustainable energy and climate change. Postgraduate courses are available in renewable energy engineering, sustainable energy systems and energy futures, offering a useful entry point for graduates with a non-relevant degree. Such courses can also help develop particular interests, and may help to secure a more senior position.

Courses that offer an industrial placement can be of particular value as they give you a chance to get a foothold in the industry and offer great networking opportunities. Seek advice from people already working in the industry when you are choosing courses, and ask about the destinations of previous students of any courses you are considering.

As most employers look for candidates with experience, relevant work experience, gained through vacation or sandwich placements, is advantageous. Experience gained through voluntary work can also be very helpful. Even though a lot of work experience opportunities are unpaid, they provide essential experience when competing for jobs in a competitive market. Becoming a student member of relevant societies, institutes or charities will not only increase your knowledge of the sector and show commitment to potential employers, but also provide opportunities to network and make useful contacts. It is also important to keep track of developments and changes in the sector.

Companies with graduate schemes (typically oil companies and power generation companies) advertise through university careers services, and in numerous careers publications aimed at graduates (for example *Inside Careers* guides, *Target* guides and the *Prospects Directory*). Some will also attend careers fairs and other on-campus events as part of the milk round, particularly at those institutions with relevant courses. Smaller specialist companies are more likely to advertise specific vacancies. Some will use recruitment agencies for this and others will advertise directly in the specialist press and through industry websites. A speculative approach may also be worthwhile.

## Getting On: Career Prospects

Opportunities for career progression and the possible directions that your career might take will vary according to the type of organization and the sector that you are working in. In well-established sectors, such as wind power, more traditional, clearly structured career progression through the ranks of a large company is more likely to be possible. In contrast, working in a small, specialist company could mean you are the only engineer and are responsible for building a team around you as the business grows. Self-employment and freelance work are possible in energy or environmental consultancy.

There is a huge range of salaries at senior level in the industry – typically £30k–£60k after 10–15 years.

### 2.3.4 Water Engineer

Water engineer is a generic title given to engineers who specialize in water-based projects. They generally deal with the provision of clean water, disposal of waste water and sewage, and prevention of flood damage. Asset management plays a major part in a water engineer's job. This involves repairing, maintaining and building structures that control water resources, for example, sea defence walls, pumping stations and reservoirs. Engineers have to constantly address new challenges and problems which are caused by global warming, ageing infrastructure, population growth and higher quality standards.

A water engineer can expect to undertake a range of activities, including both technical and non-technical tasks. The exact mix will depend on the seniority of the post, its location (office- or site-based), and the employment sector. There are, for example, differences between working in water supply or treatment and working in flood prevention, although many general engineering functions apply across the board. Tasks most relevant to those with a mechanical engineering background typically involve designing overall schemes, such as sewer improvement schemes or flood defence programmes, and associated structures, such as pumping stations, pipework and earthworks (the scale of the design may range from an initial outline to a full, detailed design).

The range of typical starting salaries is £20k–£27k. New MEng graduates can typically expect a higher starting salary than entrants with a BEng. Larger companies in the private sector tend to pay salaries at the upper end of the range; posts with a high level of management responsibility attract salaries above this level. Remuneration is often negotiable and rarely specified in job advertisements. Public sector salaries are more transparent and generally lower; public sector employers may offer a benefits package to compensate for this and these posts may require less geographical mobility.

Unsocial working hours may be a feature of some jobs, especially when deadlines are looming. Site work in particular may involve long hours as well as travel. Additional time is commonly expected of engineers and is paid by some companies. Staff with operational responsibilities or flood monitoring duties may be on a call-out rota for out-of-hours emergencies.

Jobs are generally available across the UK but mainly in London and the South East (since London's water and sewerage network systems are old and are constantly in need of attention). Geographical mobility is an important factor with some employers. Travel within the working day to visit sites is common. Absence from home at night is occasional, mainly associated with training courses or emergency call-outs. The extent of overseas travel depends on the type of employer; assignments ranging from a few weeks to a couple of years are possible in the private sector.

## Getting In: Entry

There is a growing preference by employers for the MEng degree. For undergraduate students, an MSc in a specialized area such as water engineering, hydrology or hydrogeology may be advantageous.

Relevant vacation work or an industry placement year can improve your chances, along with a demonstrable interest in water engineering, environmental issues or public health evidenced through academic projects or water-related modules undertaken as part of your degree/ postgraduate study.

National companies such as Thames Water and Yorkshire Water operate graduate training schemes and are active on many university campuses. Most employers have closing dates or offer jobs well in advance of a start date, so it is advisable to make applications early in your final year of study. Smaller companies are more likely to advertise specific vacancies. Vacancies are often advertised through recruitment agencies, especially those specializing in maritime industries. A speculative approach can be useful, in particular for arranging work experience. Recruitment agencies are now quite active in advertising posts for both permanent and temporary staff. Large national agencies are the most likely to be used by water companies.

## Getting On: Career Prospects

Most employers expect and encourage graduates to work towards full membership of the relevant professional body, though it may be a lesser requirement in contracting firms. In addition, you may work towards registration as a Chartered Engineer (CEng) or Incorporated Engineer (IEng), depending on your academic qualifications. You must also acquire competence in a range of skills beyond the purely technical, take increasing responsibility for your work and demonstrate professional commitment over a period of at least four years.

You could be expected to take on considerable responsibilities quickly, such as managing a project, controlling large-cost budgets or supervising a team of new graduates. There is plenty of scope to progress to management positions, particularly if you are prepared to study for further business-related qualifications. It is also possible to reach higher salary grades through the development of further technical expertise where management responsibilities may not be required.

Movement between employers and between the private and public sectors is relatively easy once sufficient experience has been gained, particularly after you have obtained CEng registration. You can also work as a consultant or as a contractor. A willingness to move around the UK, and the world, increases your career prospects, as private firms in particular are often looking to deploy staff on overseas projects for periods ranging from weeks to years. You may choose to develop expertise in specialist areas, such as hydraulics, geotechnics, coastal engineering or dams.

## 2.4 Further Resources

- Occupational profiles for energy engineer, mechanical engineer and more
  http://www.prospects.ac.uk/links/occupations
- Industry insight into energy and utilities
  http://www.prospects.ac.uk/links/industries
- Institution of Mechanical Engineers: Energy, Environment and Sustainability industries
  http://www.imeche.org/knowledge/industries/energy-environment-and-sustainability/
- Careers in Renewable Energy
  http://www.careersinrenewables.co.uk
- Department of Energy and Climate Change
  http://www.decc.gov.uk
- Water UK
  http://www.water.org.uk/
- Cogent: The Sector Skills Council for Chemicals, Pharmaceuticals, Nuclear, Oil and Gas, Petroleum and Polymers
  http://www.cogent-ssc.com/
- Organization of the Petroleum Exporting Countries (OPEC)
  http://www.opec.org
- Engineering Council
  http://www.engc.org.uk/
- International Energy Agency (IEA)
  http://www.iea.org/

# 3
# Aerospace and Defence

**Figure 3.1**   A group of Norwegian Bell 412 helicopters take part in a military exercise. Taken from: http://en.wikipedia.org/wiki/File:Norwegian_military_Bell_412SP_ helicopters.jpg. Source: United States Marine Corps

---

### At a Glance

- Engineers could be designing aviation systems, developing cutting-edge defence technology, providing technical support to troops on the front line or even contributing to the next era of space exploration.
- Employers: global aerospace and defence contractors, specialist companies, government defence agencies and the armed forces.
- A huge sector with a wide variety of opportunities for mechanical engineers.
- Starting salaries: £22k–£25k in the public sector, £22k–£30k in the private sector, £24k training salary then £29k–£32k starting salary in the armed forces.

---

## 3.1 Overview

If you want to work at the forefront of technology in a fast-moving and responsive industry with a global reach, then a career in aerospace or defence engineering could be for you.

You could be involved in designing systems for the latest aircraft, working on improving efficiency to reduce the carbon footprint of air travel, or providing vital technical know-how and support to troops in the frontline.

The UK aerospace industry is the largest in Europe. Employers include aerospace and aero-engine companies, research and development organizations, consultancies, the Civil Service and the armed forces. There are a small number of major global companies (e.g. BAE Systems, Rolls-Royce), large suppliers (e.g. Cobham, GKN) and numerous smaller firms that specialize in a particular product or service.

The global aerospace market has witnessed impressive growth over the past few years. Significant past increases in military budgets and growth in commercial and business aviation have fuelled this growth in spite of the tough economic conditions of 2008–2010. Despite the recent decline in orders for commercial jets and the recent pressure on government defence budgets, this is an industry that is well established in the UK with good career prospects. With an ever-increasing focus on reducing carbon footprints and environmental issues, the scope for research, design and testing within this segment of the sector is large.

Many areas of aerospace and defence offer more job security than most industries. This, in part, results from the long-term nature of the big and complex projects that companies here take on.

> Roles for mechanical engineers in the aerospace and defence sector:
> - aeronautical engineer in industry (this chapter);
> - mechanical engineer in government defence agencies (this chapter);
> - engineer in the Armed Forces (this chapter);
> - manufacturing engineer (see Chapter 6);
> - process engineer (see Chapter 6).

## 3.2 Typical Employers

Typical employers in the defence and aerospace sectors range from government departments to large multinational companies to small specialist organizations.

The UK aerospace industry is dominated by a few large and well-established companies which focus either on designing and manufacturing civilian aircraft (e.g. BAE Systems, Bombardier, Boeing and Airbus), on developing engines (e.g. Rolls-Royce), or aircraft and aircraft component manufacturing companies who complete contracts for civilian or military customers (e.g. Dowty, Thales). With global aerospace companies having many sites worldwide, there may be opportunities to travel between sites or spend periods of time working in different countries. Secondments to other companies working on the same or similar projects may also be possible, for example clients or subcontractor companies.

There is a great deal of overlap in expertise and technology between the aerospace industry and the defence technology sector. Many of the aerospace companies provide expertise, design and manufacturing capability to both civilian and military clients. Other opportunities for engineers wishing to work in defence in particular are found in the armed forces (Army, Navy and Royal Air Force) and in government research agencies such as the Defence Science and Technology Laboratory (Dstl) and the Defence Engineering and Science Group (DESG), both part of the UK Government Ministry of Defence (MoD).

Some mechanical engineers pursue their interest in this field within research departments in universities, which are often involved collaboratively with industrial partners. A few are employed by regulators such as the Civil Aviation Authority (CAA).

A few aeronautical engineers are also employed in civil aviation by airlines and airfreight operators, or in the growing market for maintenance, repair and overhaul of military and commercial aircraft. However, a more common route for engineers in the civil operating companies is to start as aircraft technicians and to develop to engineer status. There can also be opportunities to spend periods of time seconded to other companies working on the same or similar projects, for example client or subcontractor companies.

For those interested in contributing to the highly competitive field of space technology, the largest group of employers is the university sector. However, there are also opportunities to contribute to national and international space programmes through organizations such as NASA and the European Space Agency. Another option could be the manufacturing companies who work on space technologies such as EADS Astrium and Space X.

## 3.3 Engineering Roles Specific to the Aerospace and Defence Sector

### 3.3.1 Aeronautical Engineer in Industry

An aeronautical engineer applies scientific and technological principles to research, design, maintain, test and develop the performance of civil and military aircraft, missiles, weapons systems, satellites and space vehicles. Work is typically focused on developing high standards of safety and quality, as well as reducing system costs and, increasingly, the environmental impact of air travel. Aeronautical engineering offers a wide range of roles. Most engineers specialize in a particular area, such as research, design, testing, manufacture or maintenance. The aerospace industry is well established in the UK, and constant expansion in air travel means that there are many roles available.

Starting salaries vary greatly between employers, but are typically in the range £22k–£30k. Graduates with Master's or doctoral level qualifications often achieve slightly higher starting salaries than those graduating from undergraduate courses. Larger, more renowned employers may offer higher salaries. Salaries do not vary greatly between regions.

Core working hours are mainly nine to five, but extra hours may be required to complete projects to deadlines. Aeronautical engineers often work on an 'on-call' basis for consultation, for example if there is a request to change the priority of repairs or modifications, or in emergency investigation. In this sector there are often opportunities in research and design involving practical lab-based work in collaboration with manufacturing departments, universities, other industrial companies and experimental establishments.

Jobs are widely available in various locations in the UK and abroad. The UK has a very advanced aerospace industry that is at the forefront of technological and scientific development. Travel within a working day and absence from home at night are necessary for visits to aircraft workshops or hangars to inspect aircraft that require modifications or repairs, or to undertake other on-site work for clients. Many of the large companies in the aerospace and defence sector operate in several countries providing opportunities for work on projects based outside the UK. Overseas travel may also be required to attend courses and conferences on aeronautical engineering.

## Getting In: Entry

Although there are many aeronautical engineering degree courses in the UK, entry with a good honours degree in mechanical engineering or more general engineering science is also common. The defence and aerospace industry is well established and has close links with professional institutions such as the Institution of Mechanical Engineers (IMechE), through which a mechanical engineer working in the sector would achieve accreditation as a Chartered Engineer.

A pre-entry postgraduate qualification is desirable but not essential. A Master's degree in aeronautical/aerospace engineering is useful if your first degree is in a different subject. Postgraduate study will allow you to focus on a specific area of aeronautical engineering but on-the-job training may be preferred by employers.

Pre-entry experience is advisable. Many engineering degree courses include an industrial placement. Other experience may be gained through vacation work. You may be at a disadvantage without experience. Employers value industrial placements as an opportunity to identify potential graduate recruits. Work shadowing and networking are also very useful. Research likely employers thoroughly and be prepared to make applications for work experience several months ahead of when you are available. Read more about the different types of work experience possible, and how to secure it, in Chapter 7.

Large companies recruit graduates onto graduate training schemes that typically last two years and give the graduate exposure to a range of projects and training. Such schemes are widely advertised through university careers services, and in numerous careers publications aimed at graduates (for example *Inside Careers* guides, *Target* guides and the *Prospects Directory*). Some will also attend careers fairs and other on-campus events as part of the milk round, particularly at those institutions with relevant courses. Smaller specialist companies are more likely to advertise specific vacancies, and may look for graduates who already have some relevant experience. Some will use recruitment agencies for this, others will advertise directly in specialist press and through industry websites. A speculative approach may also be worthwhile.

Language skills may be useful because of joint ventures and the possibility of opportunities to travel to company sites outside the UK. Security clearance is required for some work related to defence technology, and nationality restrictions may apply in some cases.

---

### Engineer Profile

**Ben is an actuation systems engineer at Goodrich.**
Currently Ben is working on designing braking systems for Airbus aeroplanes which are used by airlines such as EasyJet and Emirates. Previously he worked for Jaguar Land Rover as a design engineer. Here he talks about the exciting projects that he's working on.

'Aircraft need to carry ever increasing numbers of people and they need to do so as efficiently as possible. To develop new aircraft we need a huge amount of innovation and design that involves hundreds of teams of people working together to bring their expertise in line to create a complex, intricate working whole. For example, back in 2003 I worked on Emirates first class interiors and this was a big challenge – getting a fully reclining armchair, desk, wardrobe, minibar and 23-inch TV into a 2m by 1m space. This whole cabin needed to be comfortable, accessible and easily maintainable.

'Ever since I watched *Back to the Future* as a kid and saw Doc Brown with his crazy inventions, I knew I wanted to be an inventor. So I did a degree in mechanical engineering. I am now a design engineer at Goodrich Actuation Systems and I'm working on a thrust reverser project for the Airbus A350.

'If you consider that the average weight of an aircraft is 180 tons and that comes into land at 150 mph – that's a lot of weight we are talking about! A thrust reverser is built into the side of the engine. It opens up an aperture through which the air rushing through and out the back is redirected forwards, decelerating the plane in a shorter distance, enabling us to have shorter runways. So, to go about designing something like this you've got to sit down and look at where are the logical places for all the components to actually sit and interact with each other. As I'm designing, a structure has to withstand a certain load. There will be certain areas in my design that we take material out of that saves weight. There will be areas where there will be stress concentrations and we've got to keep that weight in there or even add extra material to support the load structure. For example, if I get some information from the stress engineers saying an area is a bit of a weak point we can add extra material to reinforce it. Once we've designed the parts they go to the manufacturing guys who take the drawing and actually make that component.'

## Getting On: Career Prospects

The aerospace industry offers a range of career development opportunities including specialist engineering roles, project management, procurement and planning, teaching and consultancy. The aerospace industry has a strong track record of investing in its workforce through funding and supporting employee training, as well as encouraging membership of relevant engineering institutions, such as IMechE. Training varies between organizations but most large companies offer structured training and encourage professional/Chartered status, particularly companies with graduate employment schemes. Companies usually offer in-service training and short courses to meet specific training needs. Some employers work closely with universities to develop appropriate programmes for staff at different levels and in different specialisms. The larger companies invest not only in technical training, but also in specialist topics including leadership and project management. In smaller companies without a graduate training scheme there is less likely to be a formal programme of training and professional development. However, it can still be possible to access external training courses, and accreditation with the relevant professional body is usually still encouraged.

Much of the technology involved in this setting is cutting edge. Aeronautical engineers are often involved in highly innovative new concepts, making it essential for them to continuously develop themselves personally and professionally. Having trained as an aeronautical engineer, you may move into more specific areas such as aerodynamics, aeronautics, aviation systems engineering, design, aerostructures, composites, fatigue and damage tolerance, materials and process engineering, or thermodynamics. Some pilots, for example, originally trained as aeronautical engineers.

Salaries for engineers with three to five years' experience are typically in the range £35k–£55k, rising to £50k–£65k at senior levels.

### 3.3.2 Mechanical Engineer in Government Defence Agencies

The Ministry of Defence operates two key organizations to develop technical capability in the defence arena. These are the DESG and Dstl.

DESG employs 13,000 professional engineers and scientists working within the Ministry of Defence in the UK Civil Service to equip and support the UK Armed Forces with technology. Graduate engineers work in communication and information systems; weapons, munitions, ordnance and explosives; land systems; aerospace systems; maritime systems; nuclear systems: or estates/construction.

Dstl is one of the largest groups of scientists and engineers in public service in the country. It is an integral part of the Ministry of Defence as its adviser on defence-related science and technology. Dstl delivers defence research, specialist technical services and the ability to track global technological developments. It also supports procurement decisions, defence policy making and operations.

Other, less numerous, opportunities exist for engineers working for government organizations in the defence and aerospace sector. The Civil Service recruits engineers directly into the MoD and the Department for Business, Innovation and Skills. As a civil servant with an engineering background you could get involved in procurement for some of the world's most technologically advanced engineering projects or could be handling policy development or conveying complex technological ideas to ministers in layman's terms.

## Getting In: Entry

Both Dstl and DESG operate a graduate training scheme which gives recruits the opportunity to experience a range of projects and a comprehensive programme of training and professional development in their early career. Both organizations regularly attend careers fairs and other events at universities and nationally.

DESG holds assessment centres for candidates applying by early November in December each year, with further assessment centres for later applicants in March. Application is via an online form and online situational judgement exercise. Candidates must have (or expect) a minimum of a 2:2 degree and be British nationals or hold dual nationality of which one is British.

Dstl recruits approximately 150 graduates throughout the year, with vacancies advertised on their website as they arise. Applicants must be a

British national or dual nationality (including British). Some posts are not open to dual nationals.

Entry to other areas of the Civil Service for graduate engineers is via the Civil Service Faststream. Applications open in late September and close in late November each year. Places are competitive. Around 10,000 graduates apply annually for approximately 400 places across all government departments.

## Getting On: Career Prospects

Engineers working in defence in government agencies benefit from extensive opportunities for professional development. An individual progresses through a clearly defined organizational structure at a rate determined by their abilities, performance and wishes. The upward path through such organizations has become much more flexible in recent years. At Dstl, for example, it is possible for a technical expert to remain working in a wholly technical role throughout their advancement in the organization. Equally an individual can choose to explore advancement in a management role or swap between technical and management roles as their career progresses.

The close collaboration between government agencies and private companies working in aerospace and defence means that it is common for individuals to move between public and private sector jobs during their career.

### 3.3.3 Engineer in the Armed Forces

The UK Armed Services consists of three forces covering land, air and sea – the Army, the Royal Air Force (RAF) and the Royal Navy – all of which recruit many engineering graduates each year.

Engineers in the Army can be involved in anything from providing the latest maps and making sure the local drinking water is purified, to ensuring military tanks are in full working order. They can belong to one of four different Army corps.

1. The Royal Engineers are soldiers and combat engineers, responsible for using specialist equipment to help the military move, fight and survive.
2. Members of the Royal Electrical and Mechanical Engineers repair and maintain the large array of Army equipment, including fixed and rotary wing aircraft.
3. The Royal Signals are leaders in information technology and communications for the Army.
4. The Royal Logistics are responsible for getting the right kit to the right people in the right place at the right time.

The Navy and the RAF are also important recruiters of engineers to work on the maintenance of aircraft and ships, including their propulsion, navigation equipment, weaponry, and control and communication systems. Navy Air Engineer Officers are responsible for keeping aircraft at sea constantly ready to fly and may also work in a support and acquisition role, perhaps as a project manager, putting safety management schemes in place, or on new systems and solutions with the aerospace industry. Similar roles exist with a focus on ships, ship systems and landing craft (Navy Marine Engineer Officer), on submarines (Navy Marine Engineer Officer (submarine)), or their weapons (Navy Weapon Engineer Officer).

About a third of RAF personnel work in engineering. RAF Engineering Officers lead teams of technical specialists working on everything from aircraft systems to aircrew flying helmets. They may specialize in aero-systems (aircraft and missiles) or communications/electronics (radars and advanced communications equipment). RAF Engineering officers will start their careers learning practical leadership skills in a demanding technical environment. They will typically then move on to support, planning and acquisition roles, leading multidisciplinary teams, and using both core and advanced engineering skills to solve complex problems.

Most graduate engineers entering the armed forces will serve as Officers, attracting a training salary of around £24k and then a starting salary of £29k–£32k. Officers may be commissioned for 6–12 years depending on the force. Some types of work attract additional pay. For example, a submariner will receive a specialist daily allowance and aircrew officers qualify for specialist flying pay. The Army pays allowances for specialisms such as parachuting and explosives disposal.

Many engineers are UK-based, but you could also be positioned overseas, often in a conflict zone. You may potentially be involved in civilian projects, such as assisting in the redevelopment of war-torn cities or areas affected by natural disasters. Job rotation and relocation can be expected every two to three years. The extent of relocation, travel and family separation will depend upon the service and posting.

## Getting In: Entry

Generally speaking, you must be a UK, Commonwealth or Irish citizen and have been resident in the UK or Ireland for five years prior to entry to the armed forces, but some exceptions and restrictions apply. Age limits also apply.

Potential applicants to the Army should initially contact an Army careers adviser before applying for a commission in one of the corps. The next step is to attend an Army Officer Selection Board Briefing aimed at giving more information and assessing your suitability to move forward to the Army Officer Selection Board itself. This is an intensive three and a half days of physical tests, written tests, interviews and group tasks. The final hurdle is to pass a medical examination before successful candidates are offered a place at the Royal Military Academy Sandhurst to begin their training.

A similar selection process operates in the RAF. Applicants who pass initial stages attend an Officer and Aircrew Selection Centre. Successful completion of fitness tests, health assessments, aptitude tests and selection interviews leads to an 'offer of service' and a place on an officer training course.

Joining the Royal Navy begins with an initial conversation with an adviser and an Initial Careers Presentation. Next is the application form, which leads to the Recruit Test that covers basic English and maths, problem-solving and understanding of mechanics. Offers are made to those who also pass fitness tests, an eye test and a medical.

## Getting On: Career Prospects

The Army officer training course at the Royal Military Academy Sandhurst lasts for 11 months. This comprehensive programme includes practical skills training (fieldcraft, military skills, drill and fitness training), personal skills training (leadership, decision-making, negotiation, self-confidence, mental agility and communication), and adventurous training (which involves a course, followed by an expedition to test your ability to perform under challenging conditions).

Graduate officer recruits to the Royal Navy undertake a 28-week training period at the Britannia Royal Naval College Dartmouth, followed by four months of Common Fleet Time training at sea.

The training for technical graduates entering the RAF differs according to the specific role, but most officers begin their career with the 30-week long Initial Officer Training course at the Royal Air Force College Cranwell. This involves four weeks' preparation in fitness and self-discipline, to aid the transition from a civilian to an armed forces way of life, followed by military and leadership training.

Initial officer training is followed by a four- to six-month specialist training course at one of the Defence College of Aerospace Engineering sites to provide participants with the detailed skills and particular knowledge needed before assuming a first command.

Commissions in the armed forces are of fixed duration, typically 6–12 years with the possibility of extension. Officers are promoted to Lieutenant (Army), Flying Officer (RAF) or Sub-Lieutenant (RN) after initial training, and further promotions through the ranks are based on merit. Graduate engineers in the armed forces gain credits towards Chartered status during their training, and will normally gain the CEng qualification after working in a number of posts. There are frequently opportunities to study for Master's-level qualifications. This emphasis on gaining accreditation by the relevant professional bodies is highly valued by civilian employers and an important asset if pursuing a second career outside of the armed forces. Many opportunities exist for officers upon retirement from the armed forces because of the management and professional training and experience they gain during their period of service.

## 3.4 Further Resources

- Occupational profiles for aeronautical engineer, mechanical engineer and Armed Forces Technical Officer and more
     http://www.prospects.ac.uk/links/occupations
- Institution of Mechanical Engineers: aerospace industries
     http://www.imeche.org/knowledge/industries/aerospace/overview
- Royal Aeronautical Society
     http://www.raes.org.uk/
- Civil Aviation Authority
     http://www.caa.co.uk/
- Aerospace Defence Security (ADS) trade organization
     http://www.adsgroup.org.uk
- Defence Science and Technology Laboratory (Dstl)
     http://www.dstl.gov.uk
- Defence Engineering and Science Group (DESG)
     http://www.desg.mod.uk
- The Army
     http://www.army.mod.uk/
- The Royal Navy
     http://www.royalnavy.mod.uk/
- The Royal Air Force
     http://www.raf.mod.uk/
- The UK Civil Service
     http://www.civilservice.gov.uk

# 4

# Transport and Automotive

**Figure 4.1** St Pancras International Station, London. Taken from: http://commons.
wikimedia.org/wiki/File:StPancrasInternational-PS02.JPG. Source: Przemysław
Sakrajda

---

### At a Glance

- Engineers involved in all aspects of transport – from designing vehicles to working on infrastructure projects.
- Employers: major manufacturers and technology companies, train and marine operating companies, transport infrastructure organizations.
- A diverse and socially relevant sector with engineering expertise at its core.
- Starting salary: £22k–£27k.

## 4.1 Overview

We all use transport, but as a sector it also provides diverse and challenging career opportunities for engineers. From the elderly lady who relies on her community minibus to take her to the hospital each month, to the city commuter who takes the tube every day to get to and from work, transport can have an impact on everyone's lives – and engineers are at the core of delivering the infrastructure and vehicles needed.

The UK has an extensive transport network. Thirty-four million vehicles are carried on 400,000km of roads. The rail network accommodates 1.2 billion passenger journeys per year and there is also a large fleet of naval, merchant and passenger ships. Add into this the light rail, tram and underground networks, aviation (covered in Chapter 3) and freight operators, and the scale of the sector soon becomes apparent.

As a mechanical engineer working in the transport sector you could be designing the next generation of hybrid cars, or improving signalling on main lines, or perhaps managing the maintenance of a fleet of underground trains.

The buoyancy of the transport sector, and the automotive sector in particular, is always dependent on the strength of the UK and world economy. Even in competitive areas such a motorsport and vehicle design, opportunities are still readily available if candidates look outside Formula 1 companies or to designing 'yellow goods' such as trucks and heavy vehicles.

Careers in the aviation sector, and marine engineering careers with the Royal Navy, are covered in Chapter 3.

Roles for mechanical engineers in the transport and automotive sector:
- automotive engineer (this chapter);
- mechanical engineer in the rail industry (this chapter);
- naval architect (this chapter);
- manufacturing engineer (see Chapter 6);
- process engineer (see Chapter 6).

## 4.2 Typical Employers

The UK boasts several major car manufacturers, with activity still concentrated in the West Midlands. It also has a number of smaller producers serving specialist markets, such as sports and luxury cars and London taxis. There are, in addition, over 1,000 automotive component suppliers manufacturing in the UK, 90% of which are SMEs (small to medium-sized enterprises), with many offering high levels of expertise in specific technical areas. Graduate employers include car, commercial vehicle and motorcycle manufacturers, design houses and test laboratories, tyre manufacturers, accessory and safety equipment manufacturers, fuel and oil companies, and motorsport teams. The UK is recognized as a world leader in innovation in component manufacture and attracts considerable investment from international manufacturers. It is also the centre of the motorsport world, which employs around 50,000 full-time people in the UK. The most successful Formula 1 Grand Prix cars, every car that takes part in America's famous Indianapolis 500 motor race, Subaru, Mitsubishi and Nissan rally cars, and all Motor Sport BMWs are designed and built in the UK. Motorsport is one of the UK's major export earners with a total turnover of £1.3 billion.

The UK rail industry consists of two industry-wide organizations, Network Rail and the Association of Train Operating Companies (ATOC), twenty-four train operating companies (for example Virgin, First Group and Chiltern Trains) and a handful of freight operating companies (for example Freightliner and DB Schenker). These are supplied by a large number of rail equipment design and manufacturing companies which range from multinational multi-disciplinary companies (such as Siemens, Thales and Alstom), to large companies that specialize in, for example, developing and manufacturing rail vehicles (for example Bombardier Transport), to a myriad of companies that develop expertise in specific components. Network Rail runs, maintains and develops the tracks, signaling, bridges, tunnels, level crossings and viaducts. It also owns 2,500 stations and manages 18 key stations.

In London, Transport for London (TfL) is the integrated body responsible for the capital's transport system. Other companies are also involved in different aspects of the London Underground – perhaps best illustrated by the example of the Northern Line. This line is operated by London Underground. They control services, manage stations and drive the trains. The maintenance and upgrading of the infrastructure is the responsibility of TubeLines, owned by Transport for London. Another company, Alstom Transport, manages fleet maintenance and availability.

Opportunities for engineering graduates also exist in the marine industry. Here you might work for a ship-building company, be employed in the manufacturing industry that supplies components for vessels, or perhaps work in the public sector for the Royal Navy or Ministry of Defence.

## 4.3 Engineering Roles Specific to the Transport and Automotive Sector

### 4.3.1 Automotive Engineer

Automotive engineering is concerned with the design, development and production of vehicles and their component parts. Automotive engineers may specialize in a wide variety of areas including powertrain (body, chassis and engine systems), electronics and control systems, fuel technology and emissions, fluid mechanics, aerodynamics and thermodynamics.

Alongside an excellent knowledge of the engineering principles relating to their field, automotive engineers must understand and be able to use a range of new technologies in order to keep pace within a fast-moving and forward-thinking industry. The automotive industry sector broadly falls into two main areas: retail vehicles and motorsport.

Automotive engineers usually work in multidisciplinary teams to develop land-based vehicles. Their roles combine engineering expertise with management/leadership skills and their work directly influences a company's competitive edge and hence its profitability. Exact responsibilities depend on the particular area of specialism chosen (powertrain, fuel systems, etc.) and which of the three main stages of development an engineer works to support: design, research and development, or production.

Automotive design engineers design and produce visual interpretations of components and parts, using computer-aided design packages and paying attention to issues of safety, reliability, economy and environmental impact. They may also decide on the most appropriate materials for component production and be involved in building prototypes of components, developing test procedures, and conducting tests using software packages and physical testing methods.

Automotive engineers working in production are likely to take responsibility for individual projects, manage associated budgets, production schedules and resources (including staff), and supervise quality control. They may review and revise production processes in response to feedback from colleagues or clients, safety concerns, quality issues, etc., and liaise with suppliers and handle supply chain management issues. In addition there is likely to be an element of acting as a consultant on any subsequent issues or queries from clients, analysing and interpreting technical data for reports and presentations, and providing technical support to relevant internal departments such as sales and marketing.

Starting salaries vary between employers, and also depend on the amount of previous experience you have, but are typically in the range £23k–£30k.

Most automotive engineers work nine to five, unless special circum-stances require additional hours, and engineers working in manufacturing may occasionally work shifts. The motorsport industry, however, is not a nine to five sector, with many positions demanding weekend and out-of-hours working to support events and deadlines. The work depends on specialist equipment, which means you will be largely based at one site but travel to liaise with clients or other departments. Worldwide travel is common within the motorsport sector and there are also plenty of opportunities for automotive engineers in the retail motor sector to work overseas should they wish to. Some of the larger employers have worldwide offices or plants which can facilitate short-term secondments or overseas attachments.

Job availability is no longer geographically restricted to the traditional manufacturing areas, as many smaller companies across the UK offer specialist services to the automotive industry. Nevertheless, the majority of jobs are still found in the Midlands, except that most motorsport com-panies have their research, design engineering and production facilities in 'Motorsport Valley' in southern and central England.

## Getting In: Entry

It is very advantageous to have degree that is accredited by a professional institute, most probably the Institution of Mechanical Engineers (IMechE), as this is a precursor for assessment as a Chartered Engineer and employ-ers will regard your qualification more favourably.

Previous work experience is an advantage and many automotive companies offer placements and work experience schemes, particularly in motorsport. Experience can also be gained by volunteering – check Volunteers in Motorsport – or by getting involved in initiatives such as Formula Student. A pre-entry postgraduate qualification can be useful if it provides training or knowledge for very specialist roles.

Larger employers visit campuses to present and attend careers fairs. Most employers have closing dates or offer jobs well in advance of a start date, so it is advisable to make applications early in your final year of study. Recruitment agencies increasingly carry vacancies, particularly for contract work. Many advertise in the professional journals, particularly *Automotive Engineer* and *Professional Engineering*.

## Getting On: Career Prospects

Many graduate engineers join the automotive industry via a training scheme, typically of 12–24 months' duration. During this time trainees rotate around their chosen company in order to gain experience across a range of functions and disciplines. At the end of the scheme, trainees usually choose a specialist area in which to progress their career. Many training schemes are designed to meet the Initial Professional Development (IPD) criteria set down by the IMechE and are known as a Monitored Professional Development Scheme (MPDS). Once the MPDS is completed, the trainee is eligible to apply for registration as an Incorporated Engineer (IEng) or Chartered Engineer (CEng). Some universities' industrial placements are also accredited towards MPDS. The IMechE website lists companies who offer accredited MPDS training.

Achieving IEng or CEng status is the aim of most engineers in the automotive industry, as they are both highly regarded by employers.

As traditional engineering becomes more and more influenced by advancements in technology, it is essential that trainee engineers keep up to date with new developments and software packages. Most employers provide training in these areas and the IMechE also regularly runs industry-relevant lectures, seminars and workshops. Training can also be provided by the suppliers of specific software or machinery.

The progression route for many engineers will be from graduate or trainee level through to senior engineer and ultimately to a lead engineer role. This may take more than ten years, with responsibility for managing both staff and engineering projects increasing as your role develops. Managerial tasks may sometimes overshadow the technical and hands-on aspects of your work as you take on more senior positions. After four years in the industry, automotive engineers can expect to be earning £35k–£40k depending on their role and chartership status. Engineers in lead managerial roles can earn in excess of £50k. For engineers who are seeking higher financial rewards or opportunities to work overseas, contracting is often a popular choice once several years' experience have been gained. Contract engineers are effectively self-employed and move between shorter term projects. This provides variety and often excellent financial incentives, but lacks the stability and benefits associated with a position with a large company.

Engineers who wish to be at the forefront within the industry may choose to re-train or develop skills in emerging fields in order to stay ahead of their peers or take advantage of skills gaps. Environmentally friendly 'hybrid' vehicles (which use two forms of power source) and electric vehicles are currently the subjects of intense interest within the automotive sector, and there is a high demand for engineers with expertise in these technologies.

### 4.3.2 Mechanical Engineer in the Rail Industry

The day-to-day operation of the railway is carried out in an intensely practical engineering environment. Engineers working in the rail industry provide, run, maintain and develop the infrastructure, locomotives, carriages, trucks and physical environment in which they operate. Railway companies in the UK run a fleet of over 12,000 passenger carriages and locomotives plus freight fleets across Europe.

The rail industry offers a wide range of opportunities for graduate engineers. If your interests lie in research and development or in design, then one of the many companies who develop trains and rail-related equipment could be for you. Those who are more interested in the infrastructure of the rail network might look at career opportunities with Network Rail, or with one of the companies working on the London Underground such as TubeLines. More operational roles exist within the train operating companies themselves.

Starting salaries are typically £24k–£25k for new graduates entering train operating companies or Network Rail. The range is wider in design and manufacturing companies, typically £22k–£27k.

## Engineer Profile

Rebecca Broadbent, Mechanical Engineering PhD student researching development models for rail vibration and noise radiation.

'Most people have experienced railway noise at some point, but have you ever lived near a railway? Or been talking on your mobile on a platform when a train goes roaring past?

'Trains are seen as an environmentally sound form of transport, and are becoming faster and more reliable. However, as noise complaints are common and legal noise limits are introduced, railway noise research is an important topic. The challenge when working with sound is that noise is subjective; what is acceptable to one person is unacceptable to another. Just think what music you like and if there is ever a time when the same music has become annoying to someone else, for example, your parents!

'It was 6am on my 14th birthday when I realized that I wanted to be an engineer. This was when I cleaned my first steam engine. I was amazed at the machine that was in front of me! It was steam trains which got me interested in engineering and led me to do a degree in mechanical engineering, and now I work with modern railways as a research student and I am amazed at the progress which engineers have made in railway transport over the past 40 years, since the last steam train ran on mainline UK track.

'Being a research student means that I get to push boundaries and challenge assumptions to improve understanding of engineering issues. My aim is to contribute to the body of knowledge about railway noise generation, prediction and reduction. Recently I had great fun taking measurements of railway ballast; I had about 10 tonnes of stones! This is the great thing about my work, there is always the opportunity to get your hands dirty and get outdoors to take measurements if that is what you want to do.

'The work I am currently involved in is a three-year project; it is unlikely to directly impact the public due to its specific nature. With research you are never sure what outcome you are going to get or how your work will shape the area of research, and if you will find something which will affect industry and the general public. The research I carry out may influence future railway noise policy and it may only be of interest to specialist researchers. All I can be sure of is that I enjoy it, and that I am proud to be an engineer!'

A career in the rail industry could take you anywhere in the UK. Engineers on the Network Rail graduate scheme for example could find themselves in London, Swindon, Derby, Birmingham, York or Glasgow. Most engineers in the rail industry work nine to five, unless special circumstances require additional hours. However, track and signalling maintenance commonly take place overnight and at weekends so some anti-social hours working is likely in some roles.

## Getting In: Entry

Most employers in the rail industry who recruit graduate engineers expect you to have (or expect to achieve) a 2:2, 2:1. or better in a degree that is accredited by the IMechE.

Some work experience in the sector can be useful, but is by no means essential. Some train operating companies and manufacturers offer work placements and summer internships, the latter usually open to those in their penultimate year at university. For companies that don't offer formal work experience, a speculative approach can be worthwhile. Railway Technology website hosts a comprehensive online directory of companies in the industry, and a list of train operating companies and freight operating companies is available on the Network Rail website.

Graduate training programmes are run by Network Rail, ATOC and by many of the large manufacturers. The ATOC Professional Engineer Development Scheme (APEDS) is well-known in the industry, and several of the train operating companies also put their graduate recruits through the programme. APEDS lasts four years working on a series of self-selected placements and, for mechanical engineers, leads to Chartership with the IMechE. Closing dates are usually in mid-January prior to start dates in September. Many train operating companies do not run a graduate training scheme but instead welcome applications from recent graduates for specific vacancies that they advertise through their website. Freight operating companies also recruit in this way.

Larger employers visit campuses to present and attend careers fairs. Most employers have closing dates and offer jobs well in advance of a start date, so it is advisable to make applications early in your final year of study.

## Getting On: Career Prospects

There is a great deal of emphasis placed on the importance of achieving professional accreditation with the relevant engineering institution in the rail industry. The first few years in the industry will often consist of a series of placements or projects coupled with in-house or external training courses. At the end of the scheme, trainees usually choose a specialist area in which to progress their career.

The progression route for many engineers will be from graduate or trainee level through to senior engineer and ultimately to a lead engineer role. This may take more than ten years, with responsibility for managing both staff and engineering projects increasing as your role develops. Managerial tasks may sometimes overshadow the technical and hands-on aspects of your work as you take on more senior positions.

### 4.3.3 Naval Architect

A naval architect is a professional engineer who is responsible for the design, construction and repair of ships and boats. The vessels range from commercial ships such as tankers, passenger liners and ferries, military ships and submarines, to working boats such as trawlers and fishing boats or pleasure boats such as yachts and power boats. Naval architects also work on the design and construction of marine installations such as offshore drilling platforms. Modern engineering on this scale is a team activity conducted by professional engineers in their respective fields and disciplines. The naval architect integrates their activities and takes ultimate responsibility for the overall project, being careful to ensure that a safe, economic and seaworthy design is produced.

Naval architects have a broad range of employment opportunities, both in the UK and worldwide. They may be involved in designing ships and boats, and related components and specialist equipment. This is achieved by using complex mathematical and physical models to ensure that the ship's design is satisfactory technically and that it complies with safety regulations. Other roles focus on planning the whole build process of a vessel, managing everything from concept through to delivery of the final product. Some naval architects specialize in the risk analysis of ships and marine structures, using the regulations of classification societies and intergovernmental organizations such as the International Maritime Organization to assess aspects of design such as strength, stability and lifesaving equipment. Operational roles within shipping companies may also cover the complex area of ship and equipment procurement.

Naval architects may be employed in the private sector by ship building companies, offshore support companies, design consultants or marine equipment manufacturers. Opportunities also exist in the public sector with the Royal Navy, Ministry of Defence and the Defence Engineering and Science Group (see Chapter 4), in governmental and international maritime organizations, and in non-governmental organizations (NGOs) such as the Royal National Lifeboat Institution (RNLI). Classification societies (for example Lloyds Register), who are responsible for evaluating the safety of ships and marine structures, provide another possible employment route (see the International Association of Classification Societies (IACS) for more information).

Starting salaries are typically in the range of £22k–£28k. The actual rates depend on the employer, specialism, type of work and the geographical location.

Naval architects in the UK work mainly in or from coastal towns and cities, at shipbuilding yards or at ship repair facilities at ports around the coast. Some naval architects work on large-scale projects overseas and outside Europe. High-value overseas contracts are available for naval architects with high levels of experience, though these tend to be short term. Design activity is usually undertaken in an office environment and related experimental work may take place in laboratories, shipyards and offshore locations. Some aspects of a naval architect's work, such as in-ship inspections, ship trials and commissioning, may involve working in offshore locations or under demanding physical conditions.

Self-employment, consultancy, freelance work and contract work are sometimes possible, for example, as a small boat or yacht builder or design consultant, particularly in the small or high-speed craft sectors. Income will vary depending on your experience, specialism, the nature of the project and the hours worked.

## Getting In: Entry

This area of work is open to those from engineering and marine science backgrounds. However a degree that is specifically focused on naval architecture will increase your chances, so if you have a mechanical engineering degree without relevant modules, then gaining some relevant work experience and/or a post-graduate qualification such as a Master degree in Naval Architecture or Marine Engineering could greatly enhance your chances.

There are relatively few large companies that recruit graduates onto graduate training schemes. Those that exist are often advertised through university careers services, and in numerous careers publications aimed at graduates (for example *Inside Careers* guides, *Target* Guides and the *Prospects Directory*). Some will also attend careers fairs and other on-campus events as part of the milk round, particularly at those institutions with relevant courses. Smaller specialist companies are more likely to advertise specific vacancies. Vacancies are often advertised through recruitment agencies, especially those specializing in maritime industries. A speculative approach can be useful, in particular for arranging work experience. Go to the Society for Underwater Technology (SUT) or Ship Technology websites for a list of companies and other organizations active in the field.

Competition fluctuates according to the current state of the maritime industry. Submit applications to employers early in your final year. When contacting employers, check whether their training schemes have been accredited by the Royal Institution of Naval Architects (RINA) or whether they are prepared to support you in reaching the required standards.

## Getting On: Career Prospects

The education and training given to naval architects is designed to develop the skills needed to lead them to professional status, for which membership of RINA and registration as a Chartered Engineer (CEng) or Incorporated Engineer (IEng) are required. As a graduate naval architect you will usually spend around four years in the workplace, gaining experience and further training in design, engineering practice and management, before applying for Chartered status to become a fully-qualified naval architect. Once qualified, you may begin to specialize within a technical area or work in project management within the industry.

Senior naval architects tend to develop specific technical specialisms or strong project management experience, which can be used to further increase their management responsibility or develop a career in consultancy. Eventually they may move into senior executive positions. The range of typical salaries with experience is £35k– £75k. Naval architects with many years' experience working for large organizations can earn £75k or more within the UK.

## 4.4 Further Resources

- Occupational profiles for automotive engineer, naval architect and more
    http://www.prospects.ac.uk/links/occupations
- Institution of Mechanical Engineers (IMechE)
    http://www.imeche.org/
- Institution of Mechanical Engineers: Automobile; Combustion Engines and Fuels; Railway
    http://www.imeche.org/knowledge/industries
- International Maritime Organization
    http://www.imo.org
- Royal Institution of Naval Architects (RINA)
    http://www.rina.org.uk/

- International Association of Classification Societies (IACS)
  http://www.iacs.org.uk/
- Society for Underwater Technology (SUT)
  http://www.sut.org.uk/
- Ship Technology
  http://www.ship-technology.com/
- Volunteers in Motorsport
  http://www.volunteersinmotorsport.co.uk/
- Formula Student
  http://www.formulastudent.com/
- Automotive Engineer
  http://www.ae-plus.com/
- Professional Engineering
  http://www.profeng.com/
- Careers in rail – links to train and freight operating companies
  http://www.nationalrail.co.uk/contact/careers-in-rail.html
- Railway Technology – useful directory of railway technology companies
  http://www.railway-technology.com/
- Network Rail
  http://www.networkrail.co.uk/

# 5
# Healthcare

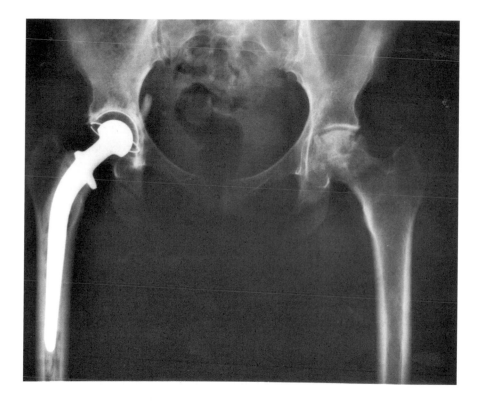

**Figure 5.1** X-Ray of an artificial hip joint. Taken from: http://en.wikipedia.org/wiki/File:746px-Hip_replacement_Image_3684-PH.jpg/ Source: NIH [Credit NIADDK, 9AO4 (Connie Raab – Contact)]

---

**At a Glance**

- Engineers involved in the design, development, installation and use of medical devices such as diagnostic imaging devices and prosthetics or in the manufacture of pharmaceuticals.
- Employers: the National Health Service (NHS) and in industry for medical device manufacturers and pharmaceutical companies.
- Rapidly evolving sector with good job prospects for graduates.
- Starting salaries: around £25k in the NHS, £20k–£30k in industry.

---

## 5.1 Overview

If you want to design hip replacements or artificial limbs, work on robotic surgery devices, or engineer mobility assistance for the disabled, then you are interested in biomedical engineering. You might also work on dialysis machines, artificial hearts or nuclear magnetic resonance (NMR) scanners.

Prospects are good: this field is growing and will grow further in future decades as medical technology advances and the population gets older. The largest employer in the UK is the NHS, but there are also opportunities in companies manufacturing healthcare devices, and in academic research and regulatory bodies, and in charities.

As a clinical engineer in the NHS you could be responsible for advising doctors on the capabilities of equipment, developing and maintaining essential medical hardware, or perhaps working directly with patients in rehabilitation centres.

In industry you could be involved in the design, development and production of instruments for monitoring health and diagnosis such as X-ray machines or ultrasound scanners for companies such as Siemens or GE Healthcare. Perhaps you'll take on a role in the design and production of implants such as replacement joints, or of metal plates to hold fractures in place while they heal, for companies such as Smith and Nephew or Corin. Or maybe you'll want to develop devices such as prosthetic limbs or mobility devices to help people to rehabilitate and live their lives more easily after suffering a physical impairment.

Alternatively you could consider a career in the pharmaceutical industry, working on the development and manufacture of drugs and vaccines. The main opportunities for mechanical engineers lie within manufacturing. You could be responsible for developing specifications, liaising with suppliers, overseeing installation, or testing and maintaining production methods and equipment. There are some large multinational companies that cover all areas of the industry. These usually recruit graduates onto training schemes that allow individuals to gain experience of a range of projects and departments. Prominent UK examples include GlaxoSmith-Kline, AstraZeneca and Pfizer. Other companies are more specialist and work in niche areas of the pharmaceutical industry. Recruitment is more likely to be for specific roles and opportunities for graduates with little experience of the industry are more limited.

Roles for mechanical engineers in the healthcare sector:
- biomedical engineer in the NHS (this chapter);
- biomedical engineer in industry (this chapter);
- manufacturing engineer (see Chapter 6);
- process engineer (see Chapter 6).

## 5.2 Typical Employers

Most mechanical engineers in the healthcare sector are employed either by the NHS or by medical-device manufacturers, although there are also opportunities within health charities, university departments and other research institutions. Well-known research units include Bath Institute of Medical Engineering (BIME) and Brunel Institute for Bioengineering (BIB). The Medical Engineering Resource Unit (MERU) designs and produces bespoke devices for individual children with disabilities, where no commercial alternatives exist. Self-employment is unlikely, although there may be scope to work as a consulting engineer or a contractor to a hospital. However, you would need to have a good network of contacts because of the collaborative nature of the work; biomedical engineers rarely work alone.

## 5.3 Engineering Roles Specific to the Healthcare Sector

### 5.3.1 Biomedical Engineer in the NHS

Biomedical engineers apply engineering principles and materials technology to healthcare. This can include: researching, designing and developing medical products, such as joint replacements or robotic surgical instruments; designing or modifying equipment for clients with special needs in a rehabilitation setting; or managing the use of clinical device in hospitals and the community.

Within the NHS, biomedical engineers are usually termed clinical engineers. Being a biomedical engineer with a mechanical engineering background is likely to involve using mathematical modelling techniques to design and improve devices, building and testing prototypes, and liaising with manufacturers, technicians and medical staff to refine the design and facilitate production. There could also be opportunities for involvement with medical staff and patients to advise on the use of the devices, planning modifications and trouble-shooting.

Most biomedical engineers are based in large hospital departments that support the full range of clinical areas. Others work within a rehabilitation unit where they play an important role in providing bespoke solutions to patients' needs for prosthetic devices, wheelchairs and a range of assistive technology. There can also be opportunities to support the research and development work of medical physics departments.

Working hours are mainly nine to five-thirty, with local variations. Those involved in research often work in a flexible environment and longer hours may be necessary at certain stages of a project. For practical reasons, safety and maintenance work on hospital equipment is likely to be performed out of hours. The workplace may be an office, laboratory, workshop, hospital or clinic, or more likely a combination of the above. Local travel within the working day may be required, for example where the job involves the regional management and maintenance of medical devices in hospitals, GP surgeries and patients' homes. Travel to meetings, conferences or exhibitions both in the UK and abroad is also possible.

Jobs are quite widely available across the UK, particularly in NHS trusts. Flexibility in preferred geographical location may be necessary both to obtain an initial training post and when seeking to move to a higher grade. NHS employees are less likely to travel abroad than private sector or research staff, who are more commonly involved in international collaboration.

The current pay for a trainee in the NHS starts at £24,831, rising to a possible £33,436 (2010 figures). The range of typical salaries for biomedical engineers working as state registered clinical scientists in the NHS is around £29k–£39k and more senior biomedical engineers in the NHS can earn up to £45k. Some may be paid on the same scale as consultants and can earn up to £90k. Actual pay rates may vary depending on the employer and location. Those working in or near London receive an additional allowance.

## Getting In: Entry

A good honours degree is essential for achieving professional recognition as a state registered clinical scientist (for clinical engineer posts in the NHS).

A pre-entry postgraduate qualification in biomedical engineering is desirable, especially for those interested in research and development. This would also improve the prospects of non-engineering graduates, or further show the commitment of those with a relevant first degree. Many Master's degree courses have opportunities for work experience in the field, which could be valuable. A Master's will not, however, reduce the number of years required to qualify as a clinical scientist.

Relevant work experience in the form of vacation work or a placement year is very helpful in getting a first job and making contacts. Voluntary or paid work with children or adults with disabilities can raise awareness of the need for products, such as specially adapted wheelchairs. Previous experience in project management, quality or design would also be useful.

NHS trainees follow a structured training programme consisting of two years' MSc and diploma study, interspersed with in-service placements. This is followed by two years in-post working in a position of responsibility while being supervised and mentored. On successful completion of an MSc, a diploma is awarded by the Institute of Physics and Engineering in Medicine (IPEM). Following two years of further training at a higher level, and acceptance of a portfolio of evidence by the Association of Clinical Scientists (ACS), graduates apply to the Health Professions Council (HPC) for state registration. This is a guarantee of competence to practise.

Competition is keen amongst graduates applying for entry to the NHS training scheme for clinical scientists, especially as the number of clinical engineer places is relatively small. Applications are made through NHS Clinical Scientists Recruitment Services. It is also worth looking out for similar posts advertised independently by local hospital trusts in the scientific press. The NHS re-advertises unfilled posts in June/July. Scotland and Northern Ireland have a limited number of training places, which are advertised through the IPEM, *New Scientist* and/or in the local press.

## Getting On: Career Prospects

Career prospects are good, as there is a slight shortage of suitably qualified and experienced applicants. A career path in the NHS has a clear structure in the early years after graduation. The main bottleneck occurs as graduates compete for a small number of pre-registration clinical scientist training posts. Beyond this point, competition for higher posts is likely to be less intense, though a willingness to relocate is important.

You could later expect to manage a department (e.g. in a hospital trust) with responsibility for medical device and technical staff across a regional area. Engineers at this level have status equivalent to medical consultants. Movement from hospital-based jobs to the healthcare industry to gain wider experience is possible.

Biomedical engineers have the opportunity to specialize in areas such as biomechanics, biomaterials, medical instrumentation or rehabilitation. Some engineers are supported in studying for a PhD. This is most common in large teaching hospitals. Others obtain fellowships with their professional body.

## 5.3.2 Biomedical Engineer in Industry

In the private sector, there is a need for engineers in companies that research and manufacture medical products such as artificial heart valves, replacement joints and monitoring equipment.

Most medical-device manufacturers are commercial companies that design, develop, manufacture and market medical devices. Biomedical engineers may be involved in all of these stages. Significant employers are multinational corporations that produce a wide range of diagnostic devices for sale to hospitals and other healthcare centres, for example GE Healthcare and Siemens. There are also many small- and medium-sized enterprises that typically specialize in a particular type of product. A web search for medical device suppliers will lead to the many online directories listing such companies. It is worth noting that the medical devices industry is highly regulated, so attention to quality assurance and design control is very important.

Working hours are mainly nine to five-thirty, but longer hours may be necessary depending on the demands of a particular project. The workplace may be an office, laboratory, workshop or manufacturing facility, or most likely a combination of these. Local travel within the working day may be required, for example where the job involves collaboration with academic researchers or suppliers, introducing the product to healthcare professionals, or involvement in clinical trials. National and international travel to meetings, conferences and exhibitions is also possible.

Self-employment is unlikely, although there may be scope to work as a consulting engineer or a contractor to a hospital. However, you would need to have a good network of contacts because of the collaborative nature of the work; biomedical engineers rarely work alone. Jobs are quite widely available across the UK.

The average starting salary for a graduate working for a large company in the pharmaceutical sector in 2009 was £26,500. This compares well with the average of £23,500 in the engineering sector (source: Association of Graduate Recruiters, 2009). Medical device manufacturing companies are diverse in size and type and also in the remuneration of their employees, however, and so graduate starting salaries will vary greatly.

Some private sector manufacturers operate internationally and may offer scope to work in Europe and beyond.

## Getting In: Entry

A good honours degree is essential for achieving professional recognition as a Chartered Engineer (CEng). Most companies on the engineering side making devices for use in surgery will employ graduates with mechanical engineering degrees. Those with qualifications in biomedical engineering are more likely to find industry jobs in companies that make implants, also known as prostheses or orthoses.

An undergraduate accredited Biomedical Engineering degree or a pre-entry postgraduate qualification in biomedical engineering is desirable but not essential. For those interested in research and development a Master's qualification and possibly a PhD would be an asset, particularly when applying for roles within specialist companies. Many MSc courses have opportunities for work experience in the field, which could be a valuable way to meet potential employers and gain insights into the industry to inform choices.

Relevant work experience in the form of vacation work or a placement year is very helpful in getting a first job and making contacts. Previous experience in project management, quality or design would also be useful.

Working in industry generally involves going into a job after your degree and working towards becoming a Chartered Engineer (CEng). Large companies recruit graduates onto graduate training schemes that typically last two years and give the graduate exposure to a range of projects and training. Such schemes are widely advertised through university careers services, and in numerous careers publications aimed at graduates (for example *Inside Careers* guides, *Target* guides and the *Prospects Directory*). Some will also attend careers fairs and other on-campus events as part of the milk round, particularly at those institutions with relevant courses. However, within the UK most medical device manufacturers (over 5,000) are SMEs; only a few are large companies. The best way to explore career prospects in a small company is to make contact directly – you won't find them on the milk round or advertising jobs at trade fairs. A good move is to request a summer placement: this gives each side a chance to see what the other is like. A good starting place for contacts is the trade association ABHI (Association of British Healthcare Industries). Smaller specialist companies are more likely to advertise specific vacancies. Some will use recruitment agencies for this; others will advertise directly in specialist press such as *New Scientist* and through industry websites.

## Getting On: Career Prospects

This is a diverse and growing industry. An ageing population coupled with the vast number of emerging medical technologies mean that opportunities to progress should be plentiful.

Engineers are usually encouraged to work towards Chartered status, and may develop their career in a predominately technical role, perhaps taking on responsibility for managing a team of engineers, or a particular suite of technologies or projects. Senior posts may offer roles in management, research, technical advice, quality assurance, production or marketing.

It is possible to move from the private sector into the public sector. Although the requirement for NHS engineers to obtain state registration before practising can make this difficult in practice.

## Engineer Profile

**Ben is a 25-year-old mechanical engineer for a medicine manufacturing company in Sussex.**

'Whilst studying mechanical engineering at university I always wanted to go into a design-based industry, maybe in defence or automotive engineering. Soon came the day where I was preparing for final exams and looking for jobs, and there were many design-based engineering jobs to choose from, and a lot of competition!

'The more I thought about it and the more interviews I went to, I decided I didn't want to sit behind a desk designing components for a final product I might not see. I then saw a position in a pharmaceutical company as a mechanical engineer which would allow me to get away from the desk and provide support to ensure reliability of equipment which makes medicines to save lives – I joined the company and have never looked back.

'Recently we've been producing antiviral- and penicillin-based drugs to combat the bird and swine flu epidemics affecting many people around the world. As an engineer it is my job to help ensure the reliability of the equipment that makes these medicines.

'Recently I purchased a thermal imaging camera to help us find hotspots on pumps and other equipment that are about to fail. The picture here shows a photo taken by the camera and the hotspots on pumps which could potentially mean eventual component failure.

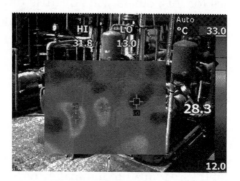

'I get to play with some cool equipment and have been involved in a variety of projects including some equipment design work, health and safety, production efficiency and energy saving projects to ensure our company produces life-saving medicines in a responsible way that doesn't damage the environment.

'Design engineering has presented me with challenging environments with lots of problem solving, and as a qualified engineer there are so many other industries that I could choose to go into that would all provide varied, challenging and active roles with great development opportunities.'

## 5.4 Further Resources

- Occupational profiles for biomedical engineer, manufacturing engineer and more
  http://www.prospects.ac.uk/links/occupations
- Institution of Mechanical Engineers: Medicine and Health
  http://www.imeche.org/knowledge/industries/medicine-and-health/overview
- National Health Service (NHS) careers information service
  http://www.nhscareers.nhs.uk/
- NHS Clinical Scientists Recruitment Services
  https://www.nhsclinicalscientists.info
- Institute of Physics and Engineering in Medicine (IPEM)
  http://www.ipem.ac.uk
- Association of Clinical Scientists (ACS)
  http://www.assclinsci.org
- Association of British Healthcare Industries
  http://www.abhi.org.uk/
- Health Professions Council (HPC)
  http://www.hpc-uk.org/
- Bath Institute of Medical Engineering
  http://www.bath.ac.uk/bime/
- Brunel Institute for Bioengineering
  http://www.brunel.ac.uk/about/acad/bib
- Medical Engineering Resource Unit
  http://www.meru.org.uk/

# 6
# Manufacturing

**Figure 6.1**  Lotus production line. Taken from: http://upload.wikimedia.org/wikipedia/commons/2/28/Final_assembly_3.jpg. Source: Brian Snelson

**At a Glance**

- Engineers involved in designing, installing, operating and developing systems to manufacture products from aeroplanes to aspirin, steelworks to sweet factories.
- Employers: a vast range, large and small, and in all industry sectors.
- An enormous sector with plentiful opportunities for good engineers.
- Starting salary: £22k–£25k.

## 6.1 Overview

Many of the UK's manufacturing industries have a heritage that extends back over centuries. Most manufacturing is now, however, a far cry from the dirty, polluting industries of the past. Today's manufacturing is a highly advanced innovative industry that operates at the cutting edge of technology.

Although the UK is the world's sixth largest manufacturer by output, some sectors within British manufacturing have for some time been characterized by declining employment and productivity. A major challenge comes from emerging economies that are able to produce goods more cheaply than the UK. The global economic slowdown and rising energy and materials costs have also affected manufacturers. Despite this, the industry is huge, employing around three million people. It accounts for 15% of UK gross domestic product (GDP) and 55% of total exports (Confederation of British Industry, 2009).

Some recent government reports (for example, *Building Britain's Future*, HM Government, 2009) have highlighted the importance of engineering and technology to the future economy of the UK. Recent initiatives aim to allow UK manufacturing to thrive by playing to its strengths of design, technology, creativity, innovation and service. To do this successfully, manufacturing needs a constant stream of well-qualified and multi-skilled graduates. Globalization also offers new opportunities with the discovery of new markets.

There are opportunities for engineering graduates in research and development, in design and in production across a vast range of sectors. Industries likely to be of particular interest to mechanical engineering graduates include aerospace (see Chapter 3), transport (see Chapter 4), biotechnology and pharmaceuticals (see Chapter 5), fast-moving consumer goods (FMCG), metals and engineered metal products, and process manufacturing.

Opportunities for engineers wanting to work in manufacturing exist throughout the UK, but traditionally some industries cluster in particular regions. For example, biotechnology and pharmaceuticals companies are concentrated on the east coast of Scotland and in South East England; motor vehicle construction takes place largely in the Midlands and the North East. Global opportunities also exist. Many companies have chosen to relocate all or part of their production lines abroad in order to compete with the cheap goods produced in low-wage economies such as India and the Far East. A key challenge for engineers working in manufacturing for UK companies is to develop processes and technological advances that enable the UK to retain (and reclaim) competitiveness.

---

Roles for mechanical engineers in the manufacturing sector:
- manufacturing engineer (this chapter);
- process engineer (this chapter).

---

## 6.2 Typical Employers

The UK manufacturing sector is incredibly diverse. An engineer working in the manufacturing sector can expect to gain employment in any industry that has an end product. Therefore, they enjoy the opportunities presented by the food and drink industry, clothing, plastics, pharmaceutical, biotechnology, oil refinery, aerospace, automotive and all types of plant and machinery manufacture. Typically, the food and drink industry together with the pharmaceutical and biotechnology industries have the largest numbers of employees.

Many manufacturing companies have become household names. In food processing, Unilever, Nestlé and Northern Foods are among the leaders, with plants throughout the UK. In pharmaceuticals, GlaxoSmith-Kline and AstraZeneca are prominent. Many automobile manufacturers and component makers have plants in the UK, including BMW, Ford, GKN, Honda, Nissan, Peugeot and Toyota. In the aerospace and defence sector, BAE Systems, Rolls-Royce, Messier-Dowty and Smiths Industries play an important role.

These big players are based in the UK but have global reach. Large manufacturing companies generally have well-developed and supported training schemes for graduate engineers and provide all-round experience and training in their chosen area of work. Responsibility is often given very early on.

In addition to these major players, most industries in manufacturing are characterized by hundreds of small- to medium-sized enterprises (SMEs). These are organizations with fewer than 250 employees. Working for a smaller company can be rewarding because as it is possible to forge a path within the company, although opportunities to try alternative departments may be limited.

Government, both local and central, also employs engineers in roles relating to manufacturing in areas such as the armed forces and utility companies.

## 6.3 Engineering Roles Specific to the Manufacturing Sector

### 6.3.1 Manufacturing Engineer

Manufacturing engineers plan, design, set up, modify, optimize and monitor manufacturing processes. They work to produce high-quality goods efficiently using the most cost-effective methods and are conscious of the environment and its protection.

Manufacturing engineers are designers, as well as analytical and creative thinkers. They can operate on their own initiative but also contribute as a team member working with engineers from various disciplines.

Manufacturing engineers also work with other professionals, in areas such as finance or health and safety.

Manufacturing engineers have the benefit of working in a wide range of areas, as the basic manufacturing principles apply to all industries. They are employed in numerous sectors, including food and drink, oil, plastics and pharmaceuticals.

The role of a manufacturing engineer varies according to the industry and the type of company but may include: designing new systems and processes for the introduction of new products or for the improvement of existing ones; working with other engineers such as chemical engineers or electrical engineers, to ensure all product and system requirements are taken into account from the initial product conception to finished result; examining and tendering for new equipment to ensure the highest quality at the best price; and organizing plant start-up and shut-down schedules to ensure minimum loss of production time and profits.

As with all professions, salary levels vary according to location, size of the organization and the nature of its business. The prevailing economic climate will also have an effect on pay. Graduate engineers can expect a starting salary in the range of £22k–£25k in the manufacturing sector. This will typically rise to £40k–£55k for senior engineers with perhaps 10–15 years' experience.

Most companies operate a shift system. Working hours may include regular unsocial hours, including weekend and evening work. You may be required to work extra hours, particularly at times when a new process is being installed and tested. Depending on what stage the manufacturing process is at, dress code will vary. The design stage will most likely be office-based, therefore a smart casual approach may be used. However, at installation stage, it may be necessary to wear full personal protection clothing and equipment. Jobs are available both in traditional industrial areas and in newer industrial estates in most parts of the UK. The company may have a sister or parent plant at other locations, opening up opportunities either nationally or internationally for travel.

## Getting In: Entry

It is advantageous to have a degree that is accredited by a professional institute, most probably the Institution of Mechanical Engineers (IMechE), as this is a precursor for assessment as a Chartered Engineer and employers will regard the qualification more favourably.

Pre-entry experience is advisable. Many engineering degree courses include an industrial placement. Other experience may be gained through vacation work. Having no experience may be a disadvantage. Employers value industrial placements as an opportunity to identify potential graduate recruits. Work shadowing and networking are also very useful. Some employers give presentations on campus or attend careers fairs. Research likely employers thoroughly and be prepared to make job applications during the first term of your final year. Look for specialist recruitment shows attended by graduate recruiters.

Large companies recruit graduates onto graduate training schemes that typically last two years and give the graduate exposure to a range of projects and training. Such schemes are widely advertised through university careers services, and in numerous careers publications aimed at graduates (for example *Inside Careers* guides, *Target* guides and the *Prospects Directory*). Some will also attend careers fairs and other on-campus events as part of the milk round, particularly at those institutions with relevant courses. Smaller specialist companies are more likely to advertise specific vacancies, and may look for graduates who already have some relevant experience. Some will use recruitment agencies for this; others will advertise directly in specialist press and through industry websites. A speculative approach may also be worthwhile.

## Getting On: Career Prospects

Structured graduate training programmes are offered by many engineering firms to new engineering graduates. These programmes often allow graduates to gain experience in different departments of the company, while gaining an understanding and appreciation of the day-to-day running of the organization. Engineers are encouraged to join their relevant professional body. Many graduate training programmes aim to assist and support new graduates to work towards gaining Chartered Engineer status. As a Chartered Engineer, you gain the designation CEng which is an internationally recognized qualification. The professional bodies, such as IMechE, also run their own continuing professional development (CPD) training programmes throughout the year.

Career development will, however, depend on the individual and their aspirations. It will also depend on the support structure in place in the organization. The organizational culture, size and nature of the business will also influence career development.

Before achieving Chartered status, the graduate or junior engineer will gain experience on numerous manufacturing processes, from the design stage through to installation. Ideally they will be exposed to the various departments within the organization and experience in-house and on-site work experience. Many engineers will progress to another company, gaining more experience and exposure to different manufacturing industries.

Some engineers will become project managers or some will specialize in a particular area and concentrate on research and development. Some may specialize to become environmental engineers or health and safety experts.

## 6.3.2 Process Engineer

A process or manufacturing systems engineer works as part of a team to design, install, monitor and develop all systems affecting the manufacturing cycle of a product. The aim is to develop and maintain efficient manufacturing systems, producing the maximum volume of high-quality product at the lowest cost and in the shortest time.

Process engineers work to integrate the entire manufacturing process, from production and supply through to sales. A systematic approach to money, methods, materials and technology across traditional departmental boundaries is required. The latest computer technology is used to provide a systematic approach to manufacturing. If you like systems and see yourself as someone who prefers to see a job through all its stages, rather than as a specialist who deals with one particular stage in the process, manufacturing systems engineering could prove very attractive.

Rather than specializing in one particular part of the process, manufacturing systems engineers are responsible for seeing a job through all of its stages. Typical work activities include: designing the layout of the plant; designing, developing and installing plant control systems; liaising with designers, researchers and engineering consultants; attending production meetings, forecasting production requirements and costs; producing maintenance schedules; testing systems and dealing with faults; reviewing results and meeting with managers to discuss methods of improving the productivity of existing systems; investigating ways in which the latest technology could improve the productivity rate of the manufacturing system and sourcing new suppliers of industrial equipment; and conducting safety tests and removing potential hazards including investigating environmental hazards.

## Getting In: Entry

Although there are specific undergraduate courses available in manufacturing systems, it is also common for mechanical engineering graduates to enter this profession. A pre-entry postgraduate qualification is not needed, but having a further qualification in a subject such as manufacturing, manufacturing systems engineering and production, or control engineering may be advantageous. In addition to the core of your engineering training, the job requires a broad and specialist knowledge of a range of engineering and operations management subjects. Increasingly, postgraduate courses are aiming to fill out the breadth of technical knowledge needed to appreciate computer-aided design, plant design and control, marketability, forecasting, production scheduling, distribution and customer feedback.

However, even with a postgraduate qualification, direct entry at the level of manufacturing systems engineer is very unlikely. Employers prefer employees to have several years' experience within an industrial environment building up their technical ability and interpersonal skills before moving into this position. Opportunities for experienced entrants are normally excellent as the coordinating and communication aspects of the job benefit from a mature approach. The demand for graduates with a suitable background and experience normally exceeds supply.

## Getting On: Career Prospects

Employers normally offer new entrants structured training through graduate training schemes, which normally last around two years. Entrants into the role of manufacturing systems engineer will typically have had prior experience of a number of different roles within the manufacturing process. Employers will, therefore, expect you to possess broad technical knowledge. They will encourage the development of knowledge and skills through on-the-job training, supplemented by short, specialist engineering courses.

Career development largely depends on the size, type and activity of your employer. For example, the company may wish to expand, increase its productivity or change/increase the range of products it manufactures.

Typically, as your career progresses, you will be expected to take on a more active role in the non-technical aspects of the job. For example, you could be negotiating prices of materials and working with plant and senior managers to improve the manufacturing efficiency. There is also training, supervising and assessing the work of trainee systems engineers and Engineering Technicians. You are likely to take part in training courses designed to improve your people and team-building skills.

Acquiring Chartered Engineer (CEng) status is a significant step in career progression. Chartered status will enable you to gain the recognition of your peers and prove that you have met a UK and international standard of experience and knowledge in the engineering profession. It will also provide you with the opportunity to network with other engineers, exchange ideas and take part in specialist industry events.

Opportunities to progress into engineering or manufacturing management, or into general management, are normally available. An increasing number of consultancies specialize in manufacturing, so subsequent movement into this field of work is a real possibility.

---

### Engineer Profile

**Hisham Fyyaz is a graduate trainee engineer for SMS Mevac UK Ltd.**
'You may have heard of Corus – it's the largest steelmaker in the UK. I work for a company that is half owned by them called SMS Mevac. We are internationally active in plant construction and mechanical engineering relating to the processing of steel, non-ferrous metals and plastics. We are specialists in secondary metallurgy.

'You're probably wondering, what does *that* mean?!! Using modern technology the steel production is divided into two steps: a primary and secondary step which makes steelmaking much more economical. This is what we do.

'The primary step is the production of a basic melt; after oxidation (addition of oxygen) this melt is tapped directly into a ladle. A ladle is a container used to transport and pour out molten metals.

'Secondary metallurgy takes place exclusively in the ladle and consists of all further measures required to improve the mechanical and chemical properties to produce high-grade steel.

'The end users of the higher grade steel come from a vast range of industries from aeroplanes, cars, housing, consumer products, energy and power generation, packaging, rail and security to defence.

'I work in the Mechanical Project Engineering Department where we look at all the mechanical aspects of the plant from design and manufacture through to the installation and commission. The work is incredibly exciting since the stuff we are designing on paper is done through applying engineering principles such as thermodynamics, stress and computer-aided design (CAD) learnt at school and university. This design work is then transformed into physical entities, structures, pipework, valve racks.

'So the things I design with a few modifications along the way will potentially end up on the steel plants. And the plant will then produce steel to make lots of different things such as the steel bracing used in buildings, the panels on the body of your car, even what your washing machine is made of! In this way, my work benefits every aspect of our modern lives.'

## 6.4 Further Resources

- Occupational profiles for process engineer, manufacturing engineer and more
  http://www.prospects.ac.uk/links/occupations
- Industry insight: Manufacturing
  http://www.prospects.ac.uk/links/manufacturingsb
- Institution of Mechanical Engineers: Manufacturing
  http://www.imeche.org/knowledge/industries/manufacturing/overview

# 7
# Getting In: Work Experience, Applications and Interviews

This chapter is about the three most important steps towards getting a job: getting experience before you apply, preparing an effective application and the interview or assessment centre.

## 7.1 Getting Experience

For many engineering roles, having some relevant work experience will greatly enhance your chances of securing your chosen job. In some sectors, such as healthcare, it can be virtually essential. Work experience is not just about adding to your curriculum vitae (CV). It is also an important part of your decision-making process and a great a way to explore the reality of a role, an organization and the industry. The ability to talk knowledgably about a sector, with real insight, is also a great way to demonstrate real interest and motivation for a role.

Sector-specific advice on how to get some work experience is given in the relevant chapters. As a general guide, however, the most common ways for students and recent graduates to gain experience are through the following.

## 7.1.1 Sandwich and Industrial Placements

Many university engineering courses allow students to take time out from
their study to work in industry for 6–12 months – usually after the second
year of their course. New graduates who want to gain some relevant
experience before applying for permanent positions are also sometimes
considered for these roles. Large companies with established graduate
schemes dominate opportunities for industrial placements.

| Sandwich and industrial placements | |
|---|---|
| *Closing dates:* | All year round |
| *How to apply:* | Via company websites |
| *Finding opportunities:* | University careers services and careers fairs |
| | Websites such as Gradcracker, Milkround.com, Inside Careers, Rate my placement, Target Jobs and Prospects |

## 7.1.2 Vacation Internships

Summer vacation placements or internships typically last 8–12 weeks
during the summer vacation before your final year as an undergraduate.
However, they may occasionally be open to new graduates prior to a gap
year. Large companies with established graduate schemes dominate opportunities for vacation internships.

| Vacation internships | |
|---|---|
| *Closing dates:* | November to March (most in February and March) |
| *How to apply:* | Via company websites |
| *Finding opportunities:* | University careers services and careers fairs |
| | Websites such as Gradcracker, Milkround.com, Inside Careers, Rate my placement, Target Jobs and Prospects |

## 7.1.3 Other Advertised Vacation Work

Many temporary and summer jobs are taken by students, and these too can help demonstrate skills and other attributes necessary to being successful later at gaining graduate entry. Often the nature of such work is below graduate level, but such employment can also give a real insight into a sector.

| Other advertised vacation work | |
|---|---|
| *Closing dates:* | Ongoing, but typically 1–3 months before start date |
| *How to apply:* | Usually CV and covering letter in response to advertisement |
| *Finding opportunities:* | University careers service vacancy databases, local press, local job centres, temping agencies |

## 7.1.4 Speculative Applications

In some sectors of engineering (e.g. medical engineering), very few organizations run structured internships. In other fields there are many more able candidates than opportunities, or you may not be eligible for advertised opportunities – if, for example, you have already graduated. In such cases it makes sense to be proactive, and actively to contact organizations to seek an opportunity to gain work experience with them.

| Speculative applications | |
|---|---|
| *Closing dates:* | Approach organizations about 3 months before your proposed start date |
| *How to apply:* | CV and covering letter |
| *Finding opportunities:* | Local business directories, company listings on professional and trade institution websites, university careers service websites and alumni networks |

## 7.1.5 Work Shadowing

Arranging a short period of shadowing experience, or just arranging to meet someone in a career field that interests you in order to get an insight into that work, is also a valuable way to demonstrate to future employers a commitment to a career field. Use existing networks of contacts such as your family and friends, alumni networks at your university, your lecturers and other members of your department to arrange initial conversations.

## 7.2 CVs and Applications

A CV and covering letter is still the most frequently asked-for method of applying for jobs. Think of them as your opportunity to market yourself, set out how you meet the criteria for the role and demonstrate your interest and motivation in the job and organization.

## 7.2.1 Notes on Example CV

1. Large, clear name – no need to say 'Curriculum Vitae'.
2. Compact contact details, appropriate email address (fluffybunnies@email.com is unprofessional!).
3. Nationality and an indication of your work permit status is important, no date of birth.
4. Mention projects and relevant degree modules.
5. An alternative format for displaying GCSE results could be '8 GCSEs (3A*, 1A, 3B, 1C).
6. Reverse chronological – the most recent and relevant activities first.
7. Give evidence of the required competences. Use bullet points and describe your achievements and responsibilities. Avoid just listing skills.

# Alex Martin ①

② 24 Any Street, Birmingham, B15 2TT, UK. Email: alex.martin@email.com Tel: 0777 777 777. Nationality: UK ③

## Education

| | |
|---|---|
| 2007–2011 | **University of Birmingham  MEng Mechanical Engineering – expected 2.1**<br>Modules included: Advanced mechanical systems; Engineering maths;<br>Thermofluids of advanced power systems; Mechanical design. ④<br>Final year research project:<br>*Thermal modeling of lightweight motors for electrical vehicles.* |
| 2000–2007 | **Anytown High School**<br>▪ A levels: Physics (A), Mathematics (B), Design &Technology (A). AS level:<br>German (B)<br>▪ GCSEs: Mathematics (A), Science (A\*A\*), Design & Technology (A\*), English<br>Language (B), English Literature (B), German (B), History (C) ⑤ |

## Work Experience

| | |
|---|---|
| Summer 2009 | **Industrial Placement, Rolls-Royce** ⑥<br>▪ 10-week placement in Noise Engineering.<br>▪ Collected and analysed data from engine rigs.<br>▪ Designed spreadsheets and macros. |
| Summer 2008 | **Placement with EPP Technologies**<br>▪ Worked in busy workshop, learning new techniques for quality control.<br>▪ Followed procedures accurately, and kept detailed records on database. ⑦ |
| 2006–2007 | **Part-time crew-member, McDonalds**<br>▪ Duties involved running the kitchen at times under substantial pressure, taking<br>orders and providing customer service at counter. |

## Positions of responsibility

| | |
|---|---|
| 2010–present | **President of the Student Engineering Society**<br>▪ Lead committee responsible for organizing seminars and social events.<br>▪ Responsible for budget allocation and some public speaking. |
| 2008–2010 | **Member of Birmingham University first women's hockey team**<br>▪ Twice weekly training sessions, played in inter-university league. |

## Other activities

Member of the Institution of Mechanical Engineers

## Additional skills ⑧

**IT Applications:** good working knowledge of LabView, MATLAB, AutoCAD, MS Office.
**Programming Languages:** high level of proficiency in Java, C++, basic knowledge of SQL, Tlearn.
**Operating Systems:** Windows, Linux.
**Languages:** good conversational German.

## Referees ⑨

Dr Bloggs, Dept of Mechanical Engineering, University of Birmingham. 0121 8888888. ibloggs@bham.ac.uk
Dr Tutor, Dept of Mechanical Engineering, University of Birmingham. 0121 7777777. otutor@bham.ac.uk

8. Give an indication of your level of ability in IT and languages (for example 'basic', 'fluent', etc.).
9. Referees – not 'references'. Stating the referees' details are available on request is also common.

### 7.2.2 Three Top Tips for a Successful CV

1. **Tailor it.** The one-size fits all CV is destined to end on the reject pile. Careful targeting is far more likely to result in success than sending off many near-identical applications. Research the organization you are applying to and the role that you are applying for, and demonstrate your understanding by providing evidence that you meet the criteria they seek.

2. **Make it your unique document.** There are no rules about the headings you must use. Beyond the necessary sections giving your name, contact details, nationality and education, you should choose headings that best display your relevant experience and skills to the potential employer. To describe your experience, for example, you might use a heading such as *Work experience, Employment history, Relevant experience, Positions of responsibility, Technical/research/engineering experience, Other experience* or *Voluntary work*. And remember: evidence that you have the competences that a potential employer is looking for can come from paid work, unpaid work, work with student societies and extra-curricular activities. Other possible headings could be *Education, Qualifications, Scholarships, Awards, Publications, Presentations, Conferences/courses attended, Interests and activities, Additional skills*, or *Languages and IT*.

3. **Think about layout and formatting.** Make sure that your CV has a clear and visually appealing layout. Assign space roughly in proportion to the relevance of the content. Make it easy for a recruiter to get a feel for what you have to offer in a short time. Don't overuse different fonts, font sizes, italics and underlining as the combined effect can

make your CV look cluttered. Also – beware using shading. It can look great on a laser-printed copy, but not so good when it has been through a photocopier.

### 7.2.3 Covering Letters

You should include a covering letter whenever you send off your CV. If you are sending your application by email, the letter can form the body of your email message, with your CV added as an attachment (ideally as a PDF document or other non-editable format). Set it out like a business letter, and ideally less than one side long. Use your letter to highlight your strengths and explain why you want the job. Give the person reading your letter a compelling reason to go on and look more closely at your CV. A suggested structure is shown in Table 7.1.

**Table 7.1**   A suggested structure for a covering letter

| Covering letter | A possible structure | Details |
| --- | --- | --- |
| Opening paragraph | Introduction | Introduce yourself; explain why you are writing and where you saw the advert. |
| Main paragraph 1 | Why you? | Explain why you are well-suited to the job by referring to relevant experience, skills and knowledge – with evidence for your claims. |
| Main paragraph 2 | Why this job, organization, sector? | This paragraph needs tailoring carefully and is your chance to show how much you know about their organization and industry. Avoid repeating text from their website or other publicity material. |
| Closing paragraph | Ending | Reiterate your desire to join the organization and add a 'look forward to hearing from you' type of statement. |

## 7.2.4 Application Forms

Not all applications require a CV and covering letter. Increasing numbers of employers create their own online forms. Most follow a similar format. They typically begin with a section for you to enter your personal details, education and qualifications – these are usually straightforward factual questions. They then go on to analytical questions designed to allow you to demonstrate that you have the personal qualities, skills and experience required, and to find out about your motivation for applying. An example of this type of question is 'Give an example of when you set yourself a demanding goal and overcame obstacles to achieve it.'

## 7.2.5 Top Tips for Effective Applications

- It's obvious, but make sure you answer the question fully! Try to understand why they are asking it – relate it back to the competencies they are looking for.
- Use specific examples as evidence – don't be too general.
- Make sure that you analyse and evaluate your part in the situations you describe on the form.
- Draw evidence from many different areas of your life.
- Check your spelling, punctuation and grammar, and ask someone else to double-check them.

## 7.3 Interviews and Assessment Centres

If you are invited to an interview then the employer has clearly been impressed with your potential. An interview is an opportunity for them to find out more about you, but also a chance for you to find out more about them, and what working in that environment would be like in reality.

Thorough preparation is the key to success at interview. You need to be able to market yourself effectively and understand how your personal qualities, skills and experience make you a suitable candidate. Be prepared to talk about the evidence that shows you meet the selection criteria. It is also important to thoroughly research the employer and the sector in which they operate. Do this by reading their website and other recruitment literature, looking through the annual report, talking to any contacts (e.g. through alumni networks at your university, or personal contacts) who have knowledge of the sector or company, and understanding the current issues facing the company by reading reports in the media. Be clear why you want to work for them.

The selection process may involve a number of stages. Many organizations do first round interviews before inviting you for a second round interview. Often phone-based, first round interviews are usually around 30 minutes and often quite informal. They are usually based on what you wrote on your application and use competency-based questions. If successful, you will then be invited for further interviews and perhaps other exercises designed to demonstrate your suitability for the role. These are known as Assessment Centres, and may involve several of the activities shown in Table 7.2.

### 7.3.1 After the Interview

If you are not successful, ask for some feedback. Employers are usually happy to give feedback to candidates who reached the assessment centre stage of the process. It can be helpful to reflect on your impressions of the interview process. Perhaps keep a note of the questions you were asked and how well you felt you handled them. Consider how you might respond differently. Use an unsuccessful interview as a learning process and use what you have learnt to improve your performance in future.

**Table 7.2** Further interview activities

| | |
|---|---|
| Psychometric tests | If you have completed tests during the application process, you may be re-tested here. These can be aptitude tests looking at specific reasoning skills (numerical, spatial, verbal) or personality questionnaires looking at factors like your motivation or team-working style. |
| Group exercise | A chance for candidates to work together as a team to solve a problem or achieve a result. This could be a practical, discussion-based or problem-solving exercise. The focus is usually on how you interact and communicate with others. |
| Presentation | You may be asked in advance to prepare a presentation on, for example, part of your course you have found interesting or a project you have done. Typically 15 minutes long, with an audience of assessors and possibly other candidates. |
| Case-study exercise | These are only occasionally used by engineering recruiters. You might be asked to review and/or analyse data relating to the employer's business and either prepare a written report or discuss your ideas in an interview. |
| Interview and/or technical interview | Second round interviews are likely to be with a panel of interviewers, including someone who works in the area you want to join. They usually focus on different areas from the first round interview and may probe issues that have arisen during other exercises. For engineering roles there are likely to be a series of technical questions – these could form a separate technical interview. |

## 7.4 Further Resources

- Prospects – applications, CVs and interviews
   http://www.prospects.ac.uk/links/appsinterviews
- University careers services
   http://www.prospects.ac.uk/links/careersservices
- Gradcracker careers website for engineering students
   http://www.gradcracker.com/
- Milkround.com
   http://www.milkround.com/

- Inside Careers
  http://www.insidecareers.co.uk/
- Rate My Placement
  http://www.ratemyplacement.co.uk/
- Target Jobs
  http://targetjobs.co.uk/

# AIM HIGHER ACHIEVE MORE.

**Institution of MECHANICAL ENGINEERS**

Become an Associate member of IMechE and you can begin the journey to professional registration as an Incorporated or Chartered Engineer. With Associate membership you can:

- Boost your earning potential
- Enhance your status with the well recognised post nominal AMIMechE
- Attend regional events and network with potential employers
- Use Career Developer, the IMechE online resource for your professional development
- Receive career advice throughout your professional life

As an Associate member you get all this and more, all for less than the cost of your daily paper, and you can even spread the cost of membership into ten manageable monthly Direct Debit payments.

To find out more and to upgrade to Associate member, visit **www.imeche.org/associate**

E membership@imeche.org
T 0845 226 9191

**Improving the world through engineering**

# 8
# Getting On: Training and Qualifications

## 8.1 Introduction

In a competitive employment market, professionals must be prepared to continually add to their skills, whether through formal learning programmes or through experience and knowledge gained from colleagues and associates. For engineers this is crucial, as employers in technology and innovation industries, as well as most professional bodies, require continued professional development from their employees. As well as being vital to healthy career progression, employing companies and organizations rely on the expertise of engineers to be proven and refreshed through professional development.

For engineering graduates, the natural step in career development is to work towards becoming professionally registered as a Chartered Engineer (CEng) or an Incorporated Engineer (IEng). These are both professional titles that demonstrate to employers and peers in industry that an engineer has reached a benchmark standard of professional competency and experience.

## 8.2 What is Professional Registration?

Professional registration means that your engineering institution has assessed your knowledge, experience and ability to do your job and concluded that you meet the high standards required to hold a professional title and registration. Being professionally registered means that you can use the post-nominal EngTech, IEng or CEng after your name, which indicates your professional calibre.

### 8.2.1 EngTech

Engineering Technicians contribute to the design, development, manufacture, commissioning, decommissioning, operation or maintenance of products, equipment, processes or services. They are required to apply safe systems of working. The EngTech professional registration is open to anyone who can demonstrate the required professional competences and commitment, typically through formal vocational qualifications or substantial working experience.

### 8.2.2 IEng

Incorporated Engineers maintain and manage applications of current and developing technology and may undertake engineering design, development, manufacture, construction and operation. Incorporated Engineers are variously engaged in technical and commercial management and possess effective interpersonal skills. The IEng professional registration is open to anyone who can demonstrate the required professional competences and commitment, plus Bachelor's level learning.

### 8.2.3 CEng

Chartered Engineers develop appropriate solutions to engineering problems. They may develop and apply new technologies, promote advanced designs and design methods and introduce new and more efficient production techniques, or pioneer new engineering services and management methods. The title CEng is protected by civil law and is one of the most recognizable international engineering registrations. Chartered Engineers are variously engaged in technical and commercial leadership, and possess effective interpersonal skills. The CEng professional registration is open to anyone who can demonstrate the required professional competences and commitment, plus Master's level learning.

These definitions are set by the Engineering Council (EC), the UK regulatory body for the engineering profession, which also sets the benchmark standards that govern professional registration and holds the national register of Chartered Engineers, Incorporated Engineers and Engineering Technicians.

The Institution of Mechanical Engineers (IMechE) offers professional development programmes to its members who wish to build on their academic skills to achieve registration as a Chartered Engineer, Incorporated Engineer or Engineering Technician. As a professional engineering institution, IMechE supports its members through the process of getting professionally registered with the EC, offering the tools and opportunities for them to reach their career potential.

Many engineers working towards professional registration are also supported in this process by their employers, as engineering companies rely on their employees to maintain their reputation for expertise, experience and excellence. By employing professionally registered engineers, they command confidence in the services and products they offer. Support from companies may range from offering paid leave to complete reports, accredited monitored professional development schemes to help you work towards professional registration to covering the cost of membership to a professional institution.

## 8.3 Benefits of Professional Registration

Using EngTech, IEng or CEng after your name distinguishes you as a valuable engineer of eminent skill. The benefits of professional registration include the following:

- **Recognition for your expertise:** Being a registered engineer means that your peers in industry have assessed your abilities and acknowledged your hard work and dedication to the profession. A gold seal of approval that marks out professional expertise, most Engineering Technicians, Incorporated Engineers and Chartered Engineers remain registered through to retirement or beyond.
- **Higher earnings potential:** Time and time again, salary surveys show that professionally registered engineers earn higher salaries earlier on in their careers than their unregistered colleagues. The most recent salary survey undertaken of the engineering industry (Engineering UK 2011 report) showed that on average Engineering Technicians, Incorporated Engineers and Chartered Engineers command annual salaries of £37,000, £43,000 and £55,000, respectively. So it really does pay to be professionally registered.
- **Improved career prospects:** Many employers and their clients require evidence that the engineers they employ have the skills and competences that they value. Professional registration gives them the assurance that the holder has experience, commitment and an internationally recognized standard of competency. It has also become a prerequisite for many senior-level engineering posts, as often tendering or post-tender contract compliance requires key members of the project team to be professionally registered engineers. This means that many professionally registered engineers find it easier to gain promotion or a new job.
- **Greater professional influence:** Becoming EngTech, IEng or CEng registered is a public confirmation of your commitment to professionalism that is vouched for by a professional institution such as IMechE. Professional registration commands respect and enables engineers to reach the most senior levels in the engineering industry and business, with their engineering credentials recognized and respected around the world.
- **Professional confidence:** Professional engineers will often cite becoming registered as EngTech, IEng or CEng as one of the proudest and most significant moments in their careers and they place high value on their professionally registered status. A great sense of satisfaction

comes from knowing that their hard work, both academic and in the workplace, has been formally recognized as having met the standards of professional excellence.

*There have been a lot of highlights in my career, but I would have to say that achieving my Chartered status was a major step – it put a real rubber stamp on my career. It gave me a real sense of pride in what I'd achieved, through my formal academic education, through my training, and confirmation of my skills and experience as an engineer. It offers me a great deal nowadays in terms of projects I've worked on for different customers.*
Oliver Tomlin, Principal Engineer, Mira Ltd

## 8.4 The Professional Registration Process

Becoming professionally registered requires knowledge, experience and commitment to both the profession and to your own career growth. A large part of your personal development will be achieved by working on real-world projects within your individual business area and reflecting on those experiences. What you learn on the job can be supplemented by a range of courses, seminars and specific events designed to fill gaps in knowledge and soft skills, offered by IMechE. These provide a great opportunity to network across industries, to develop your business and interpersonal skills, as well as helping to identify your strengths and areas for development.

### 8.4.1 Eligibility

The EngTech registration is open to anyone who can demonstrate the required professional competences either through an approved qualification (typically level 3 or level 6 in Scotland) or substantial working experience.

The IEng registration is open to anyone who can demonstrate the required professional competences and has achieved Bachelor's level learning.

The CEng registration is open to anyone who can demonstrate the required professional competences and has achieved Master's level learning.

Achieving registration as IEng or CEng is particularly straightforward for mechanical engineering graduates with an IMechE-accredited Bachelor's or Master's degree, as the qualification gained has already been reviewed and accredited as meeting the academic standards set by the EC. However, an accredited qualification is not the only path to demonstrating the necessary knowledge and understanding.

Those without a Master's degree who wish to become Chartered Engineers can either take a work-based MSc to reach Master's level learning and competency for CEng, or undertake work-based learning to Master's equivalent level. The work-based learning option requires the submission of evidence to IMechE that demonstrates the Master's equivalent level of knowledge and understanding of engineering practice has been reached.

## 8.4.2 Initial Professional Development (IPD)

Initial Professional Development (IPD) describes the first few years that lead to your registration as a professional engineer. Typically, two years of work experience after graduation will be necessary to gain Incorporated status, and four years of experience necessary to become a Chartered Engineer, although these time frames are not prescriptive and everyone is assessed on a case by case basis. Throughout your IPD, you will be supported by IMechE with a range of tools and opportunities to assess and record each competency that you have accomplished.

## 8.4.3 Monitored Professional Development Scheme (MPDS)

The accredited Monitored Professional Development Scheme (MPDS) is the IMechE's accredited and quality controlled route for graduate engineers to complete their initial professional development. The aim of the scheme is to develop engineers who can contribute to a company's success by using their technical and business competence, innovation and interpersonal skills, through a structured and monitored scheme.

MPDS encompasses all relevant training and experience accumulated from graduation through a period of IPD designed to develop the competences required by a professional engineer.

All developing engineers on the scheme are assigned a Chartered Engineer or Incorporated Engineer (of any discipline) as a mentor at the start of the programme. The mentor assists the developing engineer to achieve engineering and business competences by setting and reviewing objectives. This ongoing assessment from a professionally registered engineer helps the candidate to focus their development in the right areas.

MPDS has been designed in conjunction with industry – and over 300 engineering companies run an accredited IMechE MPDS scheme for their employees – to promote high standards of engineering practice and demonstrate a highly skilled personnel base. Some accredited universities also offer MPDS to engineering students so they can begin the process early and get a head start in their career development.

### 8.4.4 Getting Registered

If you begin the journey towards professional registration with IMechE, you will be working towards the competence framework set by the EC, which is known as UK-SPEC (the UK Standard for Professional Engineering Competence). Familiarity with UK-SPEC will be key to your development. You will also have access to our online reporting system which will enable you to keep a record of your progress.

Keeping a record of your progress during the first few years after graduation will help you to:

- remember what you've done
- understand how you meet UK-SPEC
- see what areas you need to develop as a professional

If there are gaps in your knowledge or skill, you can discuss how to meet these with your employer and mentor or attend IMechE training, seminars or events where you can bridge the gap.

The online system is a crucial part of the scheme, which helps you to do all of these and meet the requirements of the scheme. You will:

- meet regularly with your mentor to discuss and plan your progress
- write a short report of your experiences every three months against UK-SPEC
- initiate an assessment of your engineering competence every year

Once you have met the required academic benchmark and fulfilled all of the competences required for the relevant level of professional registration, you will be able to submit your documents for assessment and apply for a Professional Review Interview (PRI) with a panel of peers at IMechE to discuss your achievements. They will determine if you have met the professional competences required to be registered as an Incorporated or Chartered Engineer.

## 8.4.5 Lifelong Learning

The journey doesn't end on completion of the development programme and achievement of registration as EngTech, IEng or CEng. Professionally registered engineers must demonstrate that they are committed to keeping their competence current by undertaking lifelong learning. IMechE offers opportunities ranging from courses and lectures to events, with access to a network of similarly qualified and experienced experts in each specialist field. This keeps professional engineers up to date on the latest technological developments, maintaining their knowledge and supporting their careers as they develop.

*Getting Chartered is a recognition of your capability as an engineer and is a stamp to show that you have the qualities of a trained and competent engineer.*
Chris Tudor, Technical Development Manager, Hydratight

# 9
# Further Study and Academia

**Figure 9.1**   Graduation. Taken from: http://commons.wikimedia.org/wiki/
File:Graduation_hugs.jpg. Source: Chad Miller

Every year 20–25% of those graduating from first degrees go on to some form of further study or research. Add to these the numbers who may return to study after a gap year or after some time in employment and it is likely that around a quarter of graduates of UK universities undertake a further period of study.

The choice of subjects, qualifications and institutions available as options for you to continue your study is huge, and can be bewildering. In this chapter we will first look at possible reasons, both good and not-so-good, why you might decide to take a postgraduate qualification. Then there will be an overview of the types of courses that may be available to you, in the UK and abroad. Finally we'll consider the answers to practical considerations such as how and when to apply and how to maximize your chances of securing funding for your course.

## 9.1 What is your Motivation?

You have already done, or are doing, a degree course that has involved in-depth study of your subject. What then is your motivation for doing more study? It is critical to consider why you are thinking about undertaking a course of postgraduate study.

You may simply be keen to take your degree studies a stage further, fuelled by a **passion for your chosen subject of study**. Perhaps a particular module or topic has caught your interest and you would like to find out more about it. Maybe your interests are now taking you in a slightly different direction to your first degree, or you would like the opportunity to find out more about a subject that was not really covered in your undergraduate course. If your primary motivational factor is love of your subject, that's fine, but it is worth bearing in mind that continuing your studies can be both time-consuming (one or two years for a Master's course, three or more years for a doctoral degree) and expensive, so it is worth pausing to consider the implications.

Enthusiasm for your subject may be a very positive motivation, but if possible you should also consider your longer-term career plans. Think about the impact that your postgraduate course will have on the career options that will then be open to you. Find out from potential course tutors what students who have previously studied that course have gone on to do, and contemplate whether they are the sorts of options that you could be interested in.

Whilst this does not suggest that you should consider only career-related courses, you will ideally need to give some thought as to how you will present your decision to undertake a course of further study to future employers. If you ultimately want to work in industry, it could be well worth making the effort to develop skills during your course which would interest an employer, perhaps by keeping up a range of non-academic activities that demonstrate you have more to offer than years in the library or lab.

Perhaps you see a further qualification as a way of **enhancing your career prospects**. Entry to most areas of engineering in the UK is possible without a postgraduate degree. In many cases a postgraduate could be recruited to do the same job as a first-degree graduate. Even if your choice of course is not obviously likely to enhance your career prospects, it will give you the opportunity to develop and enhance a wide range of skills such as project management, self-motivation, time management, communication skills and so on. These valuable transferable skills gained from doing postgraduate study, carefully targeted and effectively marketed in an application, can greatly enhance job applications.

A Master's degree may well be worthwhile if you wish to **become more specialist** in order to access an area of engineering that did not form part of your undergraduate degree, or one that requires a change in direction from your previous studies. For research posts in academia or in industry, a doctorate (PhD, DPhil or EngD) may be a necessary entry requirement. Advice on the value of a postgraduate degree in the different engineering sectors is included in the earlier chapters of this book.

You might like to discuss your options with potential employers at careers fairs, presentations, through your own and university networks, or with a careers adviser.

You may consider a further course of study as a way of **delaying entering the job market,** perhaps because you are uncertain of the direction you wish to take, or as a consequence of a difficult graduate labour market at times of economic downturn. However, immersing yourself in academic work without paying attention to acquiring additional transferable skills is unlikely to improve your employment prospects. Ideally, before undertaking further study, it is important to assess the long-term career implications of such a course of action.

## 9.2 Postgraduate Courses

With more than 58,000 different postgraduate courses and research opportunities in the UK alone, the choice of course and institution can be bewildering. The main types of course are summarized in Table 9.1.

Once you have an idea of the type and subject of the course you wish to take, it is time to consider which institutions to apply to. There are many course search engines online; some are listed at the end of this chapter. Narrow down your choice of institution by talking to tutors, lecturers and potential employers for recommendations. Read research journals and specialist publications to discover the institutions that are particularly active in your subject area. Contact departments to explore your options in more detail. Find out what happens to students when they have completed the course – are they entering the professions you are interested in? Your choice is also likely to be influenced by the availability of funding, the reputation of the course/institution, location, facilities and many other factors. Reading reviews of assessments into teaching quality and research quality in the Quality Assurance Agency (QAA) and in the Research Assessment Exercise (RAE) may also be helpful. Unistats also brings together reviews of institutions and courses.

If you are considering a PhD then your choice of department, and supervisor in particular, is crucial. Think about how well you would work with your potential supervisor; whether you would be working alone or

**Table 9.1**   Main types of postgraduate courses

| | |
|---|---|
| Master's (taught)<br>e.g. MSc, MA, MBA | • Detailed study of a particular aspect of your subject<br>• May be a conversion course developing your knowledge and skills in a new subject<br>• May be vocational and lead to exemptions and/or a professional qualification<br>• Typically one year full time. Flexible and part-time courses also available<br>• Usually modular, with a project/dissertation. Some offer short industrial placements |
| Master's (research)<br>e.g. MRes, MPhil, MA (by research) | • Research-based study of a particular aspect of your subject<br>• Extended project or dissertation, with some taught modules<br>• One or two years full time<br>• Often seen as preparation for a doctoral degree |
| Doctorate<br>e.g. PhD, DPhil, EngD | • Research degree resulting in a dissertation of original material worthy of presentation as a contribution to knowledge in your subject<br>• Three or more years full time<br>• Can be started immediately following a first degree, or after a Master's<br>• Often formal registration for a PhD takes place after the initial year of research and following the successful completion of a transfer report<br>• Collaborative projects between universities and industry are common in engineering. Formal schemes include: CASE awards (awards in science and engineering tied to an applied research project and involving some time within the premises of the industrial sponsor) and Knowledge Transfer Partnerships (work in industry whilst studying for a higher degree) |
| Diplomas and certificates<br>e.g. PGCE, GDL | • Vocational courses that often form a mandatory step towards a particular profession (e.g. teaching or law)<br>• Usually 9 months full-time; part-time courses take longer |

as part of a research group; what the completion rate is for PhD students in that department; and find out what the research rating of the department is (through RAE).

There are many other factors that could influence your choice of course. Location, for example, can be important. You may find it easier and financially more viable to remain where you are, as you may have accommodation and a good network of friends. The potential to study part-time or by distance learning may also influence your decision. This flexibility might, for example, make a course easier to finance, particularly if combined with part-time work. However, you will need to consider the practicalities of juggling study with work, particularly if you also have family commitments.

## 9.3 Study Outside of the UK

Most of the information in this chapter applies both to UK courses and those outside of the UK. The majority of students who have studied in the UK choose to remain here to continue their studies. However, higher education is an increasingly international marketplace and significant numbers opt to study abroad. The most common destinations are the USA and other European countries, but opportunities exist across the globe.

While the names (Master's, PhD, etc) given to postgraduate degrees often transfer well internationally, the structure of the courses can differ greatly. In the USA, for example, most postgraduate degrees involve a combination of taught courses and research. PhDs in the USA begin with taught courses, and take at least four years to complete. Education systems vary enormously from country to country and it may not always be straightforward to transfer between them. Studying in another European country can be a good way to improve your language skills or kick-off a European career. Most courses are taught in the local language, but the opportunities to study in English are increasing.

If you are considering applying for a course of postgraduate study abroad, start your research early. Deadlines may be earlier – Fulbright scholarships for study in the USA, for example, have closing dates that are 18 months prior to your proposed start date. There are many organiza-

tions that will provide information and advice on study overseas. Several
are listed at the end of this chapter.

## 9.4 Application Timeline

---

**24–18 months to go**
Start thinking about what kind of
course you might like to do and
where you might like to do it.

---

---

**18–12 months to go**
Firm up your course choices,
check application procedures
and explore funding options.

---

---

**12–6 months to go**
Apply for places and funding.
Popular courses may fill before
Christmas; most applications
open in the Autumn and close
before Easter.

---

---

**Last 6 months**
Continue with applications
and attend interviews. Firm up
funding plans and make final
choices.

---

Ideally you would be able to follow a schedule that resembles that given
above, but for many the decision to apply for postgraduate study comes
later in your final year or perhaps during a gap year after you graduate.
If you are late in applying for postgraduate study, look out for course

advertisements in the press, such as *Prospects Postgrad* (available through your careers service) or specialist journals such as *New Scientist*, or on notice boards in your department or careers service. If you don't have a place on a course by the time your studies are completed it is worthwhile to keep looking. It is still possible that late places may become available as other students firm up their plans once degree results are published. Be prepared to respond quickly when you spot an opportunity and anticipate interviews at short notice.

## 9.5 The Application Process

There is no centralized application process for postgraduate courses. You will need to apply to each university separately. Each institution produces its own application form, list of required documents and closing dates, so you will need to research these carefully for each institution you make an application to. Applications typically consist of some or all of the following:

- **a completed application form** – usually online and including personal details, education and qualifications and a detailed personal statement, admissions essay or research proposal;
- **a curriculum vitae** (CV);
- **an official transcript** – detailing your university exam results;
- **references** – usually two or three, of which at least one will be academic;
- and, in most cases, **an application fee**.

You will find that application forms often ask how you intend to finance your study because departments need to know how many of their applicants are seeking nomination for funding awards (if any) or competing for other public funding. If no funding is available through the department then state your most likely source of funding (if you have any). You do not have to have a definite source at this stage.

Closing dates for courses in the UK typically range from December through to March – but opportunities may still exist after the deadlines if courses are not full. Where it is practical to do so, you may be called for interview. The style of interview varies greatly with course and institution. Regard interviews as a two-way process – view it as an opportunity to see the department, meet potential tutors or supervisors and possibly current students. Go prepared to talk about your motivations for applying for the course and institution; the personal and academic skills you have to offer; any relevant (academic or non-academic) experience you have; your ongoing career plans; and, for research degrees, the areas of research you propose to follow. Hopefully you will have already established that your research interests fit in well with those of the department that you are applying to. Make sure that you are aware of the research interests of key academics in the department and that you can explain what your work could add to this.

You will usually hear the outcome of your application within a couple of weeks of your interview. Your ability to take up the offer of a place may depend on the funding situation, which we will move on to discuss now.

## 9.6 Funding

Funding is usually a major issue. The actual cost will vary considerably between courses, institutions and countries, and will also depend on whether you are viewed as a 'home' or international student. You will need funding to pay for tuition fees, and for your living expenses. For example, the average cost of a one-year Master's course for a UK or EU student in the UK is over £3k, with some MBA courses costing as much as £20k. You will have to find money for your daily living expenses too, which could cost in an average year at least £7k in London and £6k elsewhere.

The majority of postgraduate students get money from a range of sources, for example, money from bursaries, part-time jobs, charitable trusts, private savings and loans. Funding bodies usually have strict eligibility criteria. For example, UK Research Councils will fund only UK and

EU students, while some scholarships are available only to international students. There is no point applying unless you are sure that you do fit their requirements. Grants can cover fees and living expenses, but they are not automatic. In the first instance you should establish how students normally fund the course you wish to apply for; course administrators and graduate admissions officers should be able to help with this.

For engineers studying postgraduate courses in engineering or other technical subjects the funding picture is less gloomy than for the graduate population as a whole. The **Engineering and Physical Sciences Research Council** (EPSRC) funds doctoral programmes, and some masters courses that serve as preparation for doctorate level courses. These EPSRC studentships are allocated to specific university departments who select which candidates to make the award to. You do not need to make a separate application in order to be considered for EPSRC awards. Just make the university to which you are applying aware that you would like to be considered when you complete their application form. Research council studentships cover tuition fees, and pay a tax-free stipend for living expenses of £13,290 (2009–2010 figure).

If you are not eligible for research council funding, or if the department that you are applying to does not have an allocation of studentships, then you will have to explore other options. Possible sources of funding include the following options:

- **Research council funding**: See above.
- **Scholarships**: Most universities offer a limited number of scholarships to students studying there. They will usually have strict eligibility criteria. Check out the graduate admissions web pages or graduate prospectus of the university where you want to study for more information. Other scholarships and awards come from charitable trusts or foundations. These are organizations that administer sums of money set aside by individuals or corporations to help specific kinds of people. There is no point applying unless you are sure that you do fit their requirements. There isn't a single source of information on awards offered by charities. Your careers service may subscribe to FunderFinder, or may

hold directories such as *Educational Grants Directory, Grants Register* or the *Directory of Grant Making Trusts*. There are also a number of online sources of information (see further resources).

• **Professional and commercial awards**: Professional institutions associated with a particular subject area sometimes produce guidance on funding courses. The website Research Professional collates details of funding opportunities. Knowledge Transfer Partnerships allow graduates to be employed while working towards a postgraduate qualification.

• **Part-time work**: This may be either within the university or outside it. These options include Research and Graduate Teaching Assistantships, which are advertised in the educational press, specialist journals and on www.jobs.ac.uk. Talk to your department about teaching opportunities. At Oxford and Cambridge you may get work giving tutorials to undergraduates through colleges, and many universities employ postgraduates as demonstrators in lab classes.

• **Loans**: Many postgraduate students need to raise funding through obtaining loans from banks or from their family. Another option for some vocational courses might be a Career Development Loan.

Finding funding for postgraduate study can be difficult. You may need to be proactive in approaching numerous charitable organizations for small sums of money. Many students fund their study through a combination of small grants, part-time work and loans. Many consider taking a gap year to earn the money to fund their study.

## 9.7 Academic Careers

---

### At a Glance

Engineers involved in teaching and research in universities, including research-only posts (usually, but not always, of a limited duration) and academic grades from lecturer to professor.
Employers: universities.
Starting salary: £25k–£35k.

---

## 9.7.1 Overview

The typical career path for an academic is to follow a first degree with a PhD and then one or more 'postdocs', hoping eventually to obtain a permanent job as a university lecturer. A postdoc is a temporary post-doctoral research job, which may last between one and four years. It is usually associated with a single research project for which the university has obtained funding. Competition for permanent academic jobs is fierce.

Once appointed as a lecturer, there is a clear progression from lecturer to senior lecturer to reader (in most institutions) and to professor: however only the most successful academics are awarded professorships. Most permanent staff carry out teaching and some administration duties as well as research. Although some universities have made an effort to improve the priority of teaching, progression typically depends much more on research performance than on teaching ability.

Success in obtaining jobs and in progression in this sector at all levels depends on demonstrating that you have the skills and experience necessary to succeed. This means collecting a track record of publications, a network of contacts and experience in attracting funding for your own projects.

The UK's universities are among the strongest in the world, and academics commonly travel around the world to take up postdoctoral jobs and permanent posts in other countries. An academic career is certainly not a straightforward continuation of the undergraduate experience. Success depends as much on networking, careful management of your career and (crucially) the ability to attract funding as on sheer brilliance.

The university sector in the UK has expanded tremendously in recent years: for example, admissions have increased 44% in the ten years to 2009 (Universities and Colleges Admissions Service). However, government is unlikely to be able to fund continued expansion in the current economic conditions.

## 9.7.2 Typical Employers

There are 166 institutions of higher education in the UK, of which 116 are universities. While nearly all carry out both teaching and research, the emphasis at different institutions is different. For example, the top five research universities (rated by HEFCE, the Higher Education Funding Council for England), are allocated one-third of the total HEFCE funding for research.

Oxford, Cambridge and Imperial College are consistently rated among the leading universities in the world for engineering. They have very large and prestigious research programmes, and competition for permanent academic posts in these institutions is very intense.

A larger group of 20 research-oriented universities, known as the Russell Group, has a similar emphasis. Outside this group, other universities and institutes of higher education will have research strengths in individual subjects, but typically not across the full range of subjects.

Even in institutions oriented less strongly towards research, a strong research record is normally essential to obtain a permanent post. This is because universities compete to obtain high research ratings from HEFCE, and these ratings determine part of the universities' funding.

Engineering departments in universities are typically more outward-looking than other disciplines, with many research projects involving collaboration with industrial partners. In fact, in many cases such collaboration is essential to obtain funding. Some postdoctoral positions will involve working in a company for part of the project.

## 9.7.3 Engineering Roles in the Higher Education Sector

### Postdoctoral Researcher

A 'postdoc' is an employee on a fixed-term contract to carry out research. Postdocs are employed by a university, and usually the job will be associated with a single research project. The university will have obtained money from a government agency or industrial sponsor for the project, and this money will be used to fund the postdoc's salary. A postdoc's boss, or supervisor, will normally be an academic (lecturer, senior lecturer, reader or professor) who manages the project and who has made the proposal for funding. Nearly all postdoctoral posts require a PhD. Employers often look for those who are already experts in the technical area concerned: this experience may be gained in a PhD or in an earlier postdoc.

There has been some concern in the higher education about the poor career prospects for postdocs. Many enter postdocs with the aim of eventually obtaining a permanent academic job, but the number of such jobs which become available is much smaller than the number of postdocs applying for them. In recent years, universities have made some effort to improve the situation. Postdocs often receive training in transferable skills which, it is hoped, will prepare them for careers in industry as well as academic life. In the same way as for a PhD, the working conditions, working hours and future prospects of a postdoc depend strongly on the supervisor.

## Research Fellows

Research universities have created some posts which bridge the gap between postdocs and permanent academic jobs. These are often known as research fellows. They vary widely between different institutions, but in a typical scheme, a leading research university might employ a few individuals with a very strong research record, taking them on after their first or second postdoctoral post and aiming to develop them towards a lectureship in three or five years. There will probably be a mixture of work on research contracts managed by others, and effort to establish your own programme of work by obtaining external funding.

Other types of jobs are also advertised as research fellowships. These include posts funded by the endowments of Oxford and Cambridge colleges, which allow the holders substantial freedom to work on their own interests, and research posts which are funded by grants encompassing a broad portfolio of research at particular institutions. Ordinary postdoctoral jobs may also be advertised as 'fellowships'.

## Lecturers, Senior Lecturers and Readers

For most academics, their first permanent job is as a lecturer. Moving from temporary employment as a postdoc to a permanent job can be very difficult. Competition is very fierce, and many postdocs find themselves going from one short-term contract to another without ever finding a lectureship.

The amount of teaching work academics are asked to do varies widely between institutions and departments. In strongly research-oriented universities, it may only be a few hours per week. In these cases, though, academics are expected to maintain a strong research programme, publishing many research papers and (most importantly) obtaining grants from outside the university to fund their research.

Applying for such grants can be very time-consuming. However it is one of the most important career-building activities for academics at every stage in their careers.

Teaching duties will include designing syllabuses, setting and marking examination papers as well as giving lectures, tutorials and laboratory classes. Academics are also expected to do other duties including interviewing students for admission, management of degree programmes, facilities or departments, and pastoral support of students.

## Professors

Professor is the highest grade of academic job. Salaries vary widely depending on the prestige and resources of the institution. In universities such as the Russell Group, professors will typically have a long track record of obtaining funding for their work and publishing work with international recognition.

## Getting In: Entry

The requirements for entry become steadily more difficult as one advances through an academic career. At all stages, there are three vital ingredients to an academic's prospects for employment: publications, a network of contacts and a record of obtaining funding.

To obtain a **postdoctoral job**, a PhD is usually essential. Employers normally look for researchers who already have knowledge in an area closely related to the project. They will also want to see published work in international journals and conferences. Competition is not very severe at this stage: PhD students who have successfully completed their projects in reasonable time and published a few research papers can normally find a postdoctoral job.

Obtaining a **lectureship** is much more difficult. Because there will be many applicants for any post, only those who can offer evidence that they will succeed will be appointed. The most important type of evidence is a record of obtaining funding. This means that a vital step in an academic career is to spend time during postdoctoral contracts preparing proposals for funding. If you can show a university's appointment committee that you have participated in several successful grant applications, you have a much better chance of obtaining a permanent job. It is also vitally important to maintain a network of contacts in your specialist area, both in your own and other institutions. Personal recommendations can be very effective in making one individual stand out from the crowd.

## Getting On: Career Prospects

Further progression, within or between institutions, again depends on contacts, publications and (primarily) obtaining funding.

Universities have well-defined salary scales for academic posts. For example, a lecturer will be appointed at a particular grade on the scale, and advance by one point per year, with a corresponding increase in salary, until he or she reaches the top of the scale. Further advancement then depends upon promotion to senior lecturer. There will be conditions for this promotion which define certain levels of achievement in teaching and research.

As in other professions, high achievers can often progress swiftly up through the scales by moving between different employers.

## 9.8 Further Resources

* Target Courses
  http://www.targetcourses.co.uk/
* Hotcourses
  http://www.hotcourses.com
* Quality Assurance Agency for Higher Education
  http://www.qaa.ac.uk
* Fulbright Commission – study in the USA
  http://www.fulbright.co.uk/study-in-the-us
* Portal on learning opportunities throughout the European space (PLOTEUS)
  http://ec.europa.eu/ploteus/
* Study in Europe
  http://ec.europa.eu/education/study-in-europe/
* Universities worldwide – searchable database of more than 8,000 universities in over 200 countries
  http://www.univ.cc/
* CASE awards
  http://www.epsrc.ac.uk/funding/students/coll/icase
* Knowledge Transfer Partnerships (KTP)
  http://www.ktponline.org.uk/
* Prospects – funding
  http://www.prospects.ac.uk/links/fundstudy
* Engineering and Physical Sciences Research Council (EPSRC)
  http://www.epsrc.ac.uk/
* Gradfund – database for postgraduates seeking help with finance
  http://www.ncl.ac.uk/postgraduate/funding

- Research Professional
  http://new.researchresearch.com
- Student finance in Wales
  http://www.studentfinancewales.co.uk
- Student finance in Northern Ireland
  http://www.studentfinanceni.co.uk
- Student Awards Agency for Scotland
  http://www.saas.gov.uk